HETEROGENEOUS NETWORKS IN LTE-ADVANCED

HETEROGENEOUS NETWORKS IN LTE-ADVANCED

Joydeep Acharya, Long Gao and Sudhanshu Gaur

Hitachi America Ltd., USA

WILEY

This edition first published 2014
© 2014 John Wiley & Sons, Ltd

Registered office
John Wiley & Sons Ltd, The Atrium, Southern Gate, Chichester, West Sussex, PO19 8SQ, United Kingdom

For details of our global editorial offices, for customer services and for information about how to apply for permission to reuse the copyright material in this book please see our website at www.wiley.com.

Library of Congress Cataloguing-in-publication data has been applied for

ISBN: 978-1-1185-1186-2 (hardback)

A catalogue record for this book is available from the British Library.

ISBN: 9781118511862

Typeset in 10/12pt TimesLTStd by Laserwords Private Limited, Chennai, India

1   2014

Contents

About the Authors

Joydeep Acharya achieved his PhD degree in Electrical Engineering at Rutgers University in 2009. He is currently a staff research engineer at Hitachi America's Wireless Systems Research Lab (WSRL) where he is involved in physical layer research and standardization in LTE-Advanced. Previously, he worked as a research consultant at GS Sanyal School of Telecommunications, Indian Institute of Technology, Kharagpur on physical layer design of WCDMA. He has been participating in 3GPP RAN 1 and 2 meetings since 2009. He is the author of several IEEE conference and journal papers and inventor of several patents filed worldwide. His research topics include MIMO signal processing, base station coordination, massive MIMO, and spectrum regulation and resource allocation for wireless systems.

Long Gao achieved his BS degree at Beijing Jiaotong University, Beijing, China, in 2003 and his MS degree at Beijing University of Posts and Telecommunications, Beijing, China, in 2006, both in Electrical Engineering. He achieved his PhD degree in Electrical Engineering at Texas A&M University, College Station, TX and joined Hitachi America Ltd., Santa Clara, CA in 2010. Since then, he has been involved in 3GPP LTE/LTE-Advanced standardization activities with focus on cooperative communication and heterogeneous networks. He has published several IEEE papers and submitted several technical contributions to 3GPP RAN 1 conference. He has served as a TPC member at major IEEE conferences such as Globecom 2010–2013, and has presented tutorials on LTE-Advanced heterogeneous network at VTC 2012 and WCNC 2013.

Sudhanshu Gaur has over 10 years of research and industry experience in the field of wireless communications. He is currently the Principal Research Engineer at Hitachi America's WSRL where he leads LTE-Advanced standardization activities. Previously, he was also involved with IEEE 802.11aa standardization and contributed to Hitachi's wireless HD video system which was demonstrated at CES 2008. Prior to joining Hitachi, he attended the Georgia Institute of Technology where he achieved his PhD degree (2005) and received MS and BTech degrees from Virginia Tech (2003) and Indian Institute of Technology, Kharagpur (2000), respectively. He is a Senior Member of IEEE, has authored several peer-reviewed publications in wireless communications, and holds several patents.

Foreword

The launch of the first LTE networks in 2009 provided a dramatic increase in the data-carrying capacity of mobile communications systems. Coupled with the rising availability and capability of smartphones, this has facilitated the uptake of new services and applications, leading to an exponential growth in mobile data traffic. As consumer expectations are fuelled by these expanding possibilities, it is forecast that this exponential trend will continue for the foreseeable future, for example by a factor of 25–50 in the next 5 years.

A variety of techniques will contribute to satisfying this rapid growth in demand. New spectrum assignments will be essential, but will only provide capacity growth which is linear with respect to the amount of spectrum allocated. Radio interface technology advances, most notably in the field of multiple antennae and coordinated multipoint operation (CoMP), will play a part; these provide only incremental advances in spectral efficiency however, and certainly not the orders of magnitude that the observed trends require. The delivery of exponential capacity growth necessitates a new approach to mobile network design, in which small cells are progressively deployed to augment the capacity of the traditional macrocellular network and to offload traffic from it. Small cells uniquely have the ability to provide exponential capacity growth for data traffic.

Heterogeneous networks, comprising macrocells complemented by large numbers of small cells, are therefore becoming increasingly important. Key to their successful deployment is the development of an understanding of the characteristics of such networks; this book is therefore both timely and apposite.

Heterogeneous networks differ from traditional homogeneous macrocellular networks in some significant respects. Unlike the macrocells, which are situated by cell planning in order to provide complete coverage, small cells are typically located according to the expected density of traffic, in so-called hotspots or hotzones. This gives rise to different and potentially stronger interference conditions which need to be managed between the macrocells and small cells, as well as between the small cells themselves in dense small cell deployments. This book explains in detail the features built into LTE Releases 8–11 to control and coordinate such interference in order to ensure successful operation of heterogeneous networks.

To get the most out of small cell deployments, it is also important to understand how to optimize the association of user equipment to cells and to balance the load between the macrocells and small cells in a way that maximizes the total system capacity. This book describes the latest research being conducted into such aspects in the 3rd Generation Partnership Project (3GPP) for LTE Release 12, as well as giving some views of possible future changes to the LTE specifications to further optimize the support of small cell deployments in heterogeneous networks.

The authors are regular participants of the standardization activities at 3GPP and are therefore well equipped to explain these features and techniques. This book will be a valuable resource for anyone needing to understand how to dramatically increase the capacity of LTE networks in practice.

Professor Matthew Baker
3GPP RAN 1 Chairman 2009–2013 & Vice-Chairman 2013–present
October, 2013 Cambridge, UK

Preface

The pace of applied science trudged at a relatively slow pace since its inception, before finally exploding in the past few decades. Few technologies embody this better than the field of wireless communications. In 1895, when in a first-ever public demonstration of its kind, Acharya Jagadish Chandra Bose used millimeter waves to ring a bell remotely and ignite gunpowder, little could anyone have imagined the impact that wireless communications were destined to have in our lives. Fast forward the next hundred years and we find a technology-driven society where ubiquitous communication between human beings has been made possible by the rapid advances in wireless systems. Nowadays, we can film a high-definition video, upload it to a social networking website using a wireless network connection and receive almost instant feedback from friends all over the globe. Among all wireless technologies, we are most reliant on our cellular phones, a trend that shows no signs of abating.

A study of modern-day cellular wireless communications is not only an exercise in technology but also has social and economic aspects. Applications and services provided by cellular providers consume more bandwidth than before and we want them to be fast, reliable and affordable. According to industry forecasts, the demand for cellular broadband data will rise to unforeseen levels in the very near future. Network operators will have to fundamentally re-think the ways they operate their networks to cope with this demand. Over the past few years, it has become apparent that one important way to achieve this is to densify the network by deploying an overlay of small cells with low transmit powers and coverage over the macro coverage area. Such hybrid systems, referred to as heterogeneous networks, will see rapid proliferation and optimization in the coming years.

As with most cellular technologies, heterogeneous networks are being developed under the auspices of the Third Generation Partnership Project (3GPP), the global collaborative effort between all interested companies and organizations. 3GPP pioneers the standardization of many theoretical solutions for wireless communications problems and enables their practical implementation through a consensus-driven process after many rounds of technical discussions and demonstrations. The present standardization activities in 3GPP are mostly centered around Long-Term Evolution (LTE). To understand the trends in current and future heterogeneous networks, an understanding of LTE and of 3GPP working procedures are required.

We have been regular attendees of the 3GPP standardization meetings over the last five years. This, coupled with our background in wireless communication research and development, puts us in the perfect position to understand the nuances of the LTE standardization process. The experiences that we have gathered over the years in 3GPP have prompted us to write this book.

Standardization in 3GPP is dynamic and evolving. Unlike classical science and engineering fields, the knowledge base of 3GPP is still evolving; published literature can therefore lag behind the state of the art in the field. To the best of our knowledge, our book is one of the very few that covers topics in LTE Releases 11 and 12 (which is the latest at the time of writing) pertaining to heterogeneous networks. We have also tried to emphasize the decision-making process in 3GPP in detail, and not limit ourselves to the final outcomes. The intermediate agreements, disagreements and discussions that lead to the final consensus on adopting a specific technology often provide valuable insights about what to expect for future standards and, by extension, feature in upcoming deployments of heterogeneous networks.

This book is not a comprehensive documentation of different LTE Releases such as 11 and 12. Instead, it takes selected topics in heterogeneous networks and attempts to describe the intuition behind the myriad agreements that comprise a 3GPP standard. Indeed, the reader is encouraged to seek supplemental knowledge about the latest agreements by reading contributions and chairman's notes of the latest 3GPP meetings.

On a final note, we would like to say that writing this book has been a challenging but rewarding experience. We have been through lots of memorable times, involving late-night shifts, intense discussions, and planning sessions. We have ourselves learnt and unlearnt much in the process of writing. Our hope now is to share some of that with the reader.

Joydeep Acharya, Long Gao, Sudhanshu Gaur
Santa Clara, California

Acknowledgements

We would like to thank the many individuals and groups who, in their own ways, have contributed towards the completion of this book. First mention goes to our collaborators in 3GPP, who are actively expanding the frontiers of LTE. Specifically, for various technical discussions, we would like to thank Rakesh Tamrakar, Pekka Kyosti, Sharat Chander, Kevin Lu, Matthew Baker, Kazuaki Takeda, Bishwarup Mondal, Hidetoshi Suzuki, Xiang Yun, Nadeem Akhtar, Lars Lindbom, Ruyue Li, Satoshi Nagata, Krishna Gomadam, Weimin Xiao, and Elean Fan.

We would like to thank our managers in Hitachi America Ltd. for their support and guidance. Special thanks go to Seishi Hanaoka for his support during the writing of this book. Thanks are also due to Norihiro Suzuki, Naonobu Sukegawa, Takahiro Onai and Ryoji Takeyari. We would also like to thank our colleagues in other Hitachi divisions, in particular the Central Research Laboratory, Japan and Hitachi China Research and Development, Beijing. In particular we would like to thank Tsuyoshi Tamaki, Hitoshi Ishida, Lu Geng and Zheng Meng for numerous discussions related to the LTE standard and its deployment. We also thank Kenichi Sakamoto, Katsuhiko Tsunehara, Kenzaburo Fujishima, Rintaro Katayama, Keizo Kusaba, Shigenori Hayase, and Hirotake Ishii.

We would like to thank Narayan Parameshwar and John McKeague for sharing their vast knowledge of LTE E-UTRAN and EPC, respectively. We would like to express our gratitude to Salam Akoum and Jayanta Kumar Acharya for providing feedback on the manuscript. Thanks to Amitav Mukherjee for providing some of the figures in the small cell deployments chapter. For the initial review of the book proposal, we would like to express our thanks to Leo Razoumov, Todor Cooklev, Nilesh Mehta and Andreas Maeder. For providing valuable content for the book and helping to obtain copyright permissions we would like to thank Patrick Merias, Nitesh Patel and Keith Mallinson. For various technical discussions about the topics covered in the book we thank Jasvinder Singh and Rahul Pupala.

Last, but not the least, we would like to thank the wonderful staff at Wiley. It has been a great experience working with them as they have been helpful and friendly at each stage of the publishing, despite our irregular delivery schedule. Special thanks go to Mark Hammond, Liz Wingett, Sandra Grayson and Susan Barclay. Many thanks also to Claire Bailey and Richard Davies.

Joydeep Acharya, Long Gao, Sudhanshu Gaur

List of Acronyms

3GPP	Third Generation Partnership Project
AAS	Active Antenna Systems
ABS	Almost Blank Subframe
ACK	Acknowledgement
AE	Antenna Element
AMC	Adaptive Modulation and Coding
ANRF	Automatic Neighbor Recognition Function
AP	Antenna Port
ARQ	Automatic Repeat Request
AS	Access Stratum
BBU	Baseband Unit
BLER	Block Error Rate
BPSK	Binary Phase Shift Keying
BSC	Base Station Controller
BSR	Buffer Status Report
BTS	Base Transceiver Station
BW	Bandwidth
C-RAN	Cloud Radio Access Network
CA	Carrier Aggregation
CAPEX	Captial Expenditure
CC	Component Carrier
CCE	Control Channel Element
CCIM	Cell Clustering Interference Mitigation
CDF	Cumulative Density Function
CDM	Code Division Multiplexing
CDMA	Code Division Multiple Access
CFI	Control Format Indicator
CIF	Carrier Indicator Field
CN	Core Network
CoMP	Coordinated Multi-Point Transmission Reception
CP	Cyclic Prefix
CP	Control Plane
CQI	Channel Quality Indicator
CRC	Cyclic Redundancy Check

CRE	Cell Range Expansion
CRS	Cell-Specific Reference Signal
CRS-IC	Cell-Specific Reference Signal Interference Cancellation
CSG	Closed Subscriber Group
CSI	Channel State Information
CSS	Common Search Space
DAS	Distributed Antenna System
DC	Direct Current
DCI	Downlink Control Information
DFT	Discrete Fourier Transform
DFT-S-OFDM	DFT Spread OFDM
DL	Downlink
DMRS	Demodulation Reference Signal
DPB	Dynamic Point Blanking
DPS	Dynamic Point Selection
DSL	Digital Subscriber Line
DTCH	Dedicated Traffic Channel
DTX	Discontinuous Transmission
DwPTS	Downlink Pilot Time Slot
ECCE	Enhanced Control Channel Element
eICIC	Enhanced Inter-Cell Interference Coordination
eNB	Evolved NodeB
eNodeB	Evolved NodeB
EPC	Evolved Packet Core
EPDCCH	Enhanced Physical Downlink Control Channel
EPS	Evolved Packet System
EREG	Enhanced Resource Element Group
FD-MIMO	Full Dimension Multiple-Input Multiple-Output
FDD	Frequency Division Duplex
FDM	Frequency Division Multiplexing
FDMA	Frequency Division Multiple Access
FEC	Forward Error Correction
FeICIC	Further Enhanced Inter-Cell Interference Coordination
FFT	Fast Fourier Transform
GP	Guard Period
GSM	Global System for Mobile Communications
HARQ	Hybrid Automatic Repeat Request
HII	High-Interference Indicator
HO	Handover
HOF	Handover Failure
HSPA	High-Speed Packet Access
ICIC	Inter-Cell Interference Coordination
IDFT	Inverse Discrete Fourier Transform
IEEE	Institute of Electrical and Electronics Engineers
IFFT	Inverse Fast Fourier Transform
IMR	Interference Measurement Resource

IMS	Internet Protocol Multimedia Subsystem
IP	Internet Protocol
ISI	Inter-Symbol Interference
ISIM	Interference Suppressing Interference Mitigation
ITU	International Telecommunication Union
JP	Joint Processing
JT	Joint Transmission
LMDS	Local Multipoint Distribution Service
LMMSE	Linear Minimum Mean Square Error Estimator
LOS	Line of Sight
LTE	Long-Term Evolution
M-LWDF	Maximum-Largest Weighted Delay First
MAC	Medium Access Control
MBMS	Multimedia Broadcast Multicast Service
MBSFN	Multicast-Broadcast Single-Frequency Network
MCS	Modulation and Coding Scheme
MIB	Master Information Block
MIMO	Multiple-Input Multiple-Output
MISO	Multiple-Input Single-Output
ML	Maximum Likelihood
MLE	Medium to Large Enterprises
MME	Mobility Management Entity
MMSE	Minimum Mean Square Error Estimator
MR	Maximum Rate
MWC	Mobile World Congress
NACK	Negative Acknowledgement
NAS	Non-Access Stratum
NCT	New Carrier Type
NDI	New Data Indicator
OAM	Operations Administration and Maintenance
OCC	Orthogonal Cover Code
OCS	Operational Carrier Selection
OFDM	Orthogonal Frequency Division Multiplexing
OFDMA	Orthogonal Frequency Division Multiple Access
OI	Overload Indicator
OLLA	Outer Loop Link Adaptation
OMADM	Open Mobile Alliance Device Management
OPEX	Operating Expense
OTA	Over The Air
PAPR	Peak to Average Power Ratio
PBCH	Physical Broadcast Channel
PCell	Primary Cell
PCFICH	Physical Control Format Indicator Channel
PCI	Physical Cell Identity
PDCCH	Physical Downlink Control Channel
PDCP	Packet Data Convergence Protocol

PDN	Public Data Network
PDSCH	Physical Downlink Shared Channel
PDU	Protocol Data Unit
PF	Proportional Fair
PHICH	Physical Hybrid-ARQ Indicator Channel
PHR	Power Headroom Report
PHY	Physical Layer
PLMN	Public Land Mobile Network
PMCH	Physical Multicast Channel
PMI	Precoding Matrix Indicator
PQI	PDSCH Rate Matching and Quasi-Co-Location Indicator
PRACH	Physical Random Access Channel
PRB	Physical Resource Block
PRG	Precoding Resource Group
PS	Packet Switched
PSS	Primary Synchronization Signal
PSTN	Public Switched Telephone Network
PUCCH	Physical Uplink Control Channel
PUSCH	Physical Uplink Shared Channel
QAM	Quadrature Amplitude Modulation
QCL	Quasi Co-Location
QPSK	Quadrature Phase Shift Keying
RACH	Random Access Channel
RAN	Radio Access Network
RAR	Random Access Response
RB	Resource Block
RBG	Resource Block Groups
RE	Resource Element
REG	Resource Element Group
RF	Radio Frequency
RI	Rank Indicator
RLC	Radio Link Control
RLF	Radio Link Failure
RLM	Radio Link Monitoring
RNC	Radio Network Controller
RNTI	Radio Network Temporary Identifier
RNTP	Relative Narrowband Transmit Power
RR	Round Robin
RRC	Radio Resource Control
RRH	Remote Radio Head
RRM	Radio Resource Management
RS	Reference Signal
RSRP	Reference Signal Received Power
RSRQ	Reference Signal Received Quality
RSSI	Received Signal Strength Indicator
S-TMSI	SAE Temporary Mobile Subscriber Identity

SAE	Service Architecture Evolution
SB	Sub-Band
SC-FDMA	Single Carrier Frequency Division Multiple Access
SCell	Secondary Cell
SCTP	Stream Control Transmission Protocol
SDIM	Scheduling-Dependent Interference Mitigation
SE	Spectral Efficiency
SFBC	Space Frequency Block Code
SFN	Single Frequency Network
SIB	System Information Block
SIC	Successive Interference Cancelation
SIM	Subscriber Identification Module
SIMO	Single-Input Multiple-Output
SINR	Signal to Interference plus Noise Ratio
SISO	Single-Input Single-Output
SLNR	Signal to Leakage and Noise Ratio
SMB	Small and Medium Businesses
SME	Small and Medium Enterprises
SNR	Singal to Noise Ratio
SOHO	Small Office/Home Office
SON	Self-Organizing Networks
SPS	Semi-Persistent Scheduling
SR	Scheduling Request
SRS	Sounding Reference Signal
SSS	Secondary Synchronization Signal
SVD	Singular Value Decomposition
TA	Tracking Area
TB	Transport Block
TDD	Time Division Duplexing
TDM	Time Division Multiplexing
TDMA	Time Division Multiple Access
TM	Transmission Mode
TPC	Transmit Power Control
TPMI	Transmitted Precoding Matrix Indicator
TTI	Transmission Time Interval
TTT	Time to Trigger
UCI	Uplink Control Information
UDP	User Datagram Protocol
UE	User Equipment
UL	Uplink
ULA	Uniform Linear Array
UMTS	Universal Mobile Telecommunications System
UpPTS	Uplink Pilot Time Slot
USS	UE-Specific Search Space
UTRAN	Universal Terrestrial Radio Access Network
VoIP	Voice over Internet Protocol

VRB	Virtual Resource Block
WCDMA	Wideband Code Division Multiple Access
WiFi	Wireless Fidelity
WiMAX	Worldwide Interoperability for Microwave Access
ZC	Zadoff Chu

1

An Introduction to Heterogeneous Networks

1.1 Introduction

We have witnessed the wireless telecommunications revolution in the last decade, and it is still very much alive. Wireless technologies have enabled new services that have empowered human beings and transformed social interactions. From the days when advance booking was necessary to make a long-distance voice call, we have come to an era when we can film a high-definition video, upload it to facebook via a wireless network, and receive almost instant feedback from our friends all over the globe. The services we use today consume more bandwidth than before and we want them to be fast, reliable, and affordable.

Various market surveys all concur about the rapid proliferation in mobile broadband data. The Strategy Analytics [1] estimates that mobile data traffic grew by 100% in 2012. As shown in Figure 1.1, the data traffic is expected to increase by about 400% by 2017. The major contributors to the traffic are bandwidth-intensive real-time applications such as mobile gaming and video. In 2013, the number of mobile-connected devices exceeded the world population and by 2017 there will be about 1.4 mobile devices per capita [2].

Are mobile operators prepared to meet this enormous surge in demand, however? It is quite a formidable challenge. It has been shown that network capability using traditional macrocell-based deployments is growing at about 30% less than the demand for data [3]. The profit margins of most operators have also been decreasing globally [4]; there are two main reasons for this. The first is that the flat rate pricing policies that customers have become accustomed to prevent the mobile data revenues of an operator to scale proportionately with the increased usage of mobile broadband data. The second is the cost (capital expenditure or CAPEX versus operating expenses or OPEX) incurred as a result of setting up more base stations to provide increased capacity and coverage, connecting them via backhaul and their operation and maintenance.

To cater for the increasing data traffic and also enhance profits, the operators therefore need to fundamentally rethink methods of operating their networks. The key principle is to deliver higher capacity at a reduced cost.

Heterogeneous Networks in LTE-Advanced, First Edition. Joydeep Acharya, Long Gao and Sudhanshu Gaur.
© 2014 John Wiley & Sons, Ltd. Published 2014 by John Wiley & Sons, Ltd.
Companion Wesite: www.ltehetnet.com.

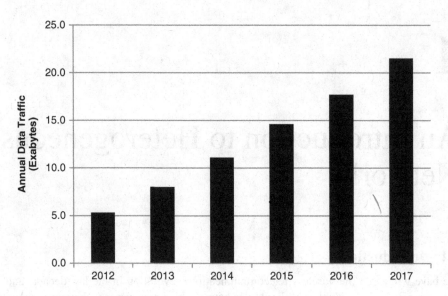

Figure 1.1 Growth forecast in annual mobile data traffic. (Source: [1]. Reproduced with permission of Strategy Analytics 2013.)

As mentioned by Mallinson [5], a 1000× increase in capacity is required to support the rising demand. High capacity can be achieved by improving spectral efficiency, employing more spectrum, and by increasing network density. The first two are related to link level enhancements where radical gains cannot be expected over current networks that are already functioning at near optimal. Instead, as Figure 1.2 from [5] shows, the major gains are expected through increasing network density by deploying an overlay network of small cells over the macro coverage area. A small cell could be an indoor femtocell or an outdoor picocell. It could be a

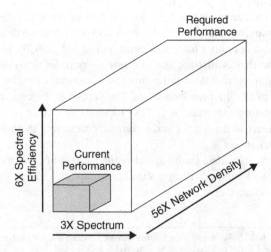

Figure 1.2 A quantitative prediction about capacity enhancement for future wireless networks

compact base station or a distributed antenna system (DAS) controlled by a central controller. The different types of small cells have low transmit power and coverage and together with the macro are referred to as *Heterogeneous Networks* or simply *HetNets*.

The reason why heterogeneous networks improve capacity is intuitive. Mobile broadband data is highly localized as the majority of current traffic is generated indoors and in hotspots such as malls and convention centers. It therefore makes sense to add capacity where it is needed by deploying an overlay of small cells in those regions of the macro coverage area which generates heavy data demand. Small cells offload data from the macro coverage area and improve frequency reuse. Additionally, they can also offer higher capacity than the macro as they can better adapt to the spatio-temporal variations in traffic by dynamic interference management techniques.

Besides improving network capacity, small cells also address the second concern of operators: cost reduction. A small cell-based heterogeneous network is much more energy efficient than a macrocell network. The power amplifier in a macrocell consumes the most energy and requires a fixed DC power supply even if there are no data being transmitted. The high transmit power requires a cooling unit which also consumes energy. The low transmit power of the small cell reduces the impact of both. It can be shown that deployment of picocells reduces overall energy consumption by 25–30%, compared to macrocell-only baseline [6]. Due to rising energy costs (electricity bills could be as high as 17% of an operator's OPEX), there is a substantial reduction in the operator's expenditure. Reducing energy consumption, a greener alternative, also helps to drive down the total carbon footprint of the operator. According to a Bell Labs study, incorporating small cells into the network can save service providers 12–53% in CAPEX and 5–10% in OPEX, depending on traffic loading.

Small cell-based heterogeneous networks are therefore being increasingly adopted by wireless operators globally. In February 2013, Informa Telecoms & Media announced at the Mobile World Congress (MWC) that, for the first time, femtocells make up the majority (56%) of all base stations deployed globally. This trend is set to continue with an 8× increase expected in the total number of deployed small cells by 2016. Clearly, small cells and heterogeneous networks play a critical role in the global mobile infrastructure.

1.2   Heterogeneous Network Deployments

As mentioned in the previous section, there are many different kinds of small cells which results in different kinds of heterogeneous networks. Each heterogeneous network has unique deployment, coverage, and capacity characteristics. The needs of the operator and their customers determine what type of heterogeneous network is optimum for a given situation. A broad classification of the different types of heterogeneous networks is provided in the following sections [7].

1.2.1   Distributed Antenna Systems

A Distributed Antenna System (DAS) consists of a network of nodes that are connected via fiber to a central processing unit. These nodes, also called DAS nodes, contain antennae/remote radio heads (RRH) that perform all the radio functionalities. The central processing unit contains the baseband transceiver module. DAS systems are usually technology agnostic, which means that the same DAS network can support multiple frequency bands and multiple technologies such as high-speed packet access (HSPA) and long-term evolution (LTE). DAS network installations usually require significant upfront capital investment and careful planned deployment by the DAS vendor. They can be deployed both outdoors and indoors and often cover large areas such as sports stadiums and convention centers.

Each DAS node is capable of supporting up to 16 frequency bands and up to 300 simultaneous connections for an area of 5000–25 000 square feet when driven by a macrocell. LTE and Worldwide Interoperability for Microwave Access (WiMAX) operators utilize an optical driver-to-fiber-to-RF (radio frequency) DAS configuration, where the optical transceiver at the central processing unit transmits the baseband signal along the fiber to remote sites where the signal is converted to RF for the first time. Other DAS solutions can be configured as RF-to-optical-to-RF systems, where the central processor converts RF signals generated by the radio transceiver of the wireless service provider to optical signals that are transported to the DAS nodes where the optical signal is converted back to RF.

The planned nature of DAS deployments leads to higher flexibility and customization according to the needs of the customer. However, it lacks the simple plug-and-play architecture that makes other small cell deployments, such as femtocells, attractive.

1.2.2 Public Access Picocells/Metrocells

As the name suggests, these small cells are deployed by operators to target specific public access areas such as railway stations and aircraft. They are open to all members of the public, although there could be a preferential grade of service for first responders such as police and public safety personnel.

The term metrocell is sometimes used to denote small cells (usually picocells) installed in a high-traffic urban area. They have less transmit power and coverage than the macrocell. Pico/metrocells are planned installations in the same way as DAS systems. Unlike DAS however, they cover a smaller area and are specific to a particular wireless access technology such as LTE or HSPA. Each picocell is capable of supporting only a few frequency bands and up to 80 simultaneous connections for an area of about 10 000–20 000 square feet. To increase coverage over a large area, an operator may install multiple picocells; they are not typically interconnected like DAS, however.

Public-access pico/metrocells have more features and are typically priced higher than consumer-grade femtocells (discussed in the following section). They will also be produced in lower numbers than femtocells. According to ABI research, the reason for this is that public-access metrocells are deployed for rugged outdoor use and need advanced features to handle simultaneous subscribers at higher RF power levels than a femtocell.

1.2.3 Consumer-Grade Femtocells

Femtocells are small stand-alone low-power nodes that are typically installed indoors. Unlike picocells, femtocell services are available only to paid subscribers who are said to form a *Closed Subscriber Group* (CSG). They support simple plug-and-play architecture and do not require professional installation by operators. The femtocells connect to the operator's network through the subscriber's internet service.

Table 1.1 Typical characteristics of femtocell-based indoor heterogeneous networks

Deployment	Coverage (ft^2)	Tx Power (mW)	Capacity (users)	Market share %	Revenue %
Residential	5000	20	8	85	70
SME/SMB	7500	100	32	15	30
Large	10 000	250	64		

Like picocells, femtocells also support a single wireless access technology. The coverage and number of supported users in a femtocell vary depending on the nature of their installation. Femtocells could be residential, also known as Small Office/Home Office (SOHO) or enterprise. Some typical values are listed in Table 1.1. It can be seen that, although enterprise femtocells have a smaller share of the market, they yield more revenue per unit than residential femtocells. Further, enterprise femtocells can be categorized into Small and Medium Businesses/Enterprises (SMB or SME), Medium to Large (MLE) or large enterprises.

1.2.4 WiFi Systems

Unlike other types of small cells, WiFi systems operate in the unlicensed band. Devices such as personal computers and smartphones can use WiFi to connect to the internet when in a residential environment. WiFi systems are inexpensive to use and install. WiFi usage is widespread in many countries and often used by businesses such as coffee shops and airports to offer free wireless access to their customers. WiFi can also integrate with an existing cellular network by offloading some of its load. The number of operators and vendors utilizing this option is steadily growing.

WiFi is based on the Institute of Electrical and Electronics Engineers (IEEE) 802.11 standards. Unlike cellular standards, the basic version of 802.11 is based on carrier sense multiple access (CSMA) technology. Latest releases of 802.11 have incorporated many quality of access (QoS) features not present in the basic CSMA version. WiFi has a range of 20–200 ft.

1.3 Features of Heterogeneous Networks

Heterogeneous networks have several features that set them apart from macro-only networks, some of which have already been alluded to in previous sections. This section presents a broad overview of some of the important features.

1.3.1 Association and Load Balancing

One of the main functions of small cells is to offload user equipment (UE) traffic from the macro. The amount of offloading depends on the criterion by which a UE associates with a base station. Downlink reference signal received power (RSRP) is the most basic criterion but this does not lead to much offloading since the transmit power of a macrocell is much greater than that of the small cell. Adding a bias to the small cell RSRP is an example of an association method that increases offloading.

The macrocell and the picocell can operate at different carrier frequencies. This is especially true for future heterogeneous networks, where small cells will be deployed in newly available spectrum in higher-frequency bands. RSRP does not capture the information about the difference in the levels of interference in the macrocell and small cell carrier frequencies. The interference information is captured in a reference signal received quality (RSRQ) based association and hence leads to better load balancing.

Association is also tied to the network load. For highly loaded systems, load balancing will distribute the UE load across all base stations uniformly. This will homogenize the inter cell interference and lead to a fair distribution of a base station's resources among all its associated UEs. For networks with low load, it may however be better to concentrate the load on a few base stations as this will reduce intercell interference from the base stations with little or no associated UEs. This principle is opposite to load balancing and is sometimes called *load shifting*.

Association can also be based on UE application or speed. For example, associating a voice user or a UE with high velocity to the macrocell will lead to better performance.

1.3.2 Interference Management

The dense deployment of small cells increases interference; unless interference mitigation techniques are applied, the gains of heterogeneous networks will not be realized. For example, when a CSG femtocell is deployed, it may interfere with the UEs that are close but cannot associate with it as they are not part of its subscriber group. Another example could be a UE that is originally connected to a macrocell but is later handed over to a picocell with a cell range expansion (CRE) bias for the purpose of load balancing. In this case, the UE will experience an interference level that is higher than the desired signal. The system performance will degrade if the intercell interference is not managed properly.

In order to coordinate the intercell interference, various techniques have been proposed and adopted in LTE Releases 8–12. Frequency-domain techniques were proposed in Release 8, where two neighboring cells can coordinate their data transmission and interference in frequency domain. Time-domain techniques and their enhancements were proposed in Releases 10 and 11, where a cell can mute some subframes to reduce its interference to its neighboring cell. In Release 12, a small cell can perform dynamic activation/deactivation based on its traffic load and interference situation. Implementation of full-dimension (FD) multiple-input-multiple-output (MIMO), where a base station can direct a narrow and focused beam directed towards the three-dimensional location of the UE, is a new area for interference management in dense heterogeneous networks. Interference can also be mitigated by coordinated multipoint transmission/reception (CoMP) technology where a macrocell and a small cell can cooperate to simultaneously serve a UE.

1.3.3 Self-Organizing Networks

Self-organizing networks (SON) are an important class of base station functionalities through which the various base stations in the network (notably the small cells) can sense their environment, coordinate with other base stations and automatically configure their parameters such as cell ID, automatic power control gains, and so on. Previously these properties formed part of the network configuration tools and processes which were configured by the operators manually. The manual methods work for a small number of homogeneous macrocell deployments but do not scale in a dense small-cell-based heterogeneous network. SONs are therefore critical to small cell deployments.

A SON optimizes network parameters for controlling interference which has a major impact on performance. It also manages the traffic load among different cells and different radio access networks and provides the user with the best possible service while maintaining an acceptable level of overall network performance. SONs reduce OPEX and thus save money for operators.

1.3.4   Mobility Management

In a cellular network, handover is performed between cells to ensure that a mobile UE is always connected to the best serving cell. The general handover process is as follows. A UE measures the signal strength of its neighboring cells. If the signal strength of a neighboring cell is higher than that of its serving cell plus an offset for a particular time period called the time to trigger (TTT), the UE will report this information to its serving cell. The serving cell then initiates the handover process.

In a homogeneous network, the handover parameters such as the handover offset and TTT are common for all cells and all UEs. However, using the same set of handover parameters for all cells/UEs may degrade the mobility performance in a heterogeneous network. It is desirable to have a cell-specific handover offset for different classes of small cells. Furthermore, for high-mobility UEs passing through a dense heterogeneous network, the normal handover process between small cells will lead to very frequent changes in the serving cell. This can be solved by associating this UE to the macrocell at all times, leading to UE-specific handover parameter optimization. Both cell-specific and UE-specific handover functionalities therefore need to be considered for heterogeneous networks.

1.4   Evolution of Cellular Technology and Standards

Performance optimization in dense small-cell heterogeneous networks is one of the most important topics in LTE. LTE-compliant small cells will see the maximum growth when compared to small cells belonging to other radio access technologies. This provided the inspiration for this book.

An evolution of various cellular standards over the years to LTE-Advanced is shown in Figure 1.3. The first generation of cellular systems, not shown in the figure, were analog systems. Second-generation cellular networks were hybrid Frequency Division Multiple Access (FDMA) and Time Division Multiple Access (TDMA) systems. Third-generation systems had Code Division Multiple Access (CDMA) as their primary radio access technology. The fourth generation of cellular networks underwent a major revolution with the introduction of Orthogonal Frequency Division Multiple Access (OFDMA) as the radio access technology and a flat, all-IP (internet protocol) core network design [8]. The first Third-Generation Partnership Project (3GPP) release with this technology was called Release 8, popularly known as LTE. Each successive release has seen gradual evolution, although there have been major leaps in technology such as the introduction of coordination among different base stations of a HetNet in LTE Release 11. A summary of some of the requirements for different 3GPP releases is provided in Table 1.2.

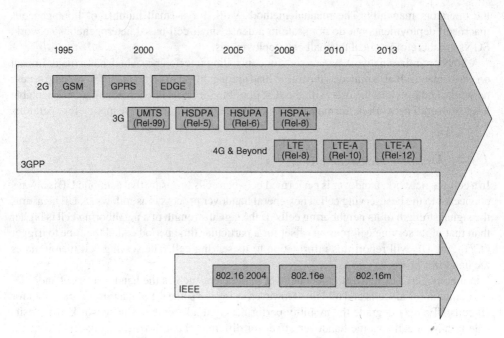

Figure 1.3 Evolution of various cellular systems

Table 1.2 Performance requirements of various 3GPP releases

		Rel-6	Rel-8	Rel-10
Downlink	Peak rate (Mbps)	14.4	100	1000
	Peak SE (bps/Hz)	3	5	30
	Cell average SE (bps/Hz)	0.53	1.6–2.1	2.6
	Cell edge SE (bps/Hz)	0.02	0.04–0.06	0.09
Uplink	Peak rate (Mbps)	11	50	500
	Peak SE (bps/Hz)	2	2.5	15
	Cell average SE (bps/Hz)	0.33	0.66–1.0	2
	Cell edge SE (bps/Hz)	0.01	0.02-0.03	0.07
System	Bandwidth (Mbps)	5	1.4–20	100
	User plane delay (ms)	–	10	10
	Control plane delay (ms)	–	100	10–50

These requirements are influenced by the guidelines of the International Telecommunications Union (ITU) which is the apex worldwide regulatory body in telecommunications. The ITU allocates spectra on a worldwide basis and decides the technologies for which the spectra are be used. The high-level requirements that these technologies should meet are also stipulated by ITU. The family of technologies that meet ITU's International Mobile Telecommunication (IMT) requirements are popularly known as 3G technologies. Similarly, LTE is an example of a technology that meets (in fact exceeds) the requirements of ITU's IMT-Advanced

Table 1.3 UE categories

UE category	Max number of DL TB bits received within a TTI	Max number of supported layers in DL	Max number of UL TB bits transmitted within a TTI	Support for 64QAM in UL
Category 1	10296	1	5160	No
Category 2	51024	2	25456	No
Category 3	102048	2	51024	No
Category 4	150752	2	51024	No
Category 5	299552	4	75376	Yes
Category 6	301504	2 or 4	51024	No
Category 7	301504	2 or 4	102048	No
Category 8	2998560	8	1497760	Yes

class of technologies (popularly known as 4G). WiMAX and 3GPP2 are other examples of technologies that meet the IMT-Advanced requirements.

As the capabilities being offered by the network changes, changes are also needed in the specifications of the user equipment (UE). Accordingly, 3GPP has introduced the notion of UE categories as listed in Table 1.3. A higher-category UE can receive and transmit data at a higher rate. It can avail the advanced features of a LTE release which a lower category UE may not be able to. For example, the categories 6–8 were added in LTE Release 10 to address the bandwidth increase due to Carrier Aggregation. UE categories thus lead to service and price differentiation.

1.4.1 3GPP Standardization Process

A major change from initial generations of cellular systems to latter releases is increased collaboration among various companies leading to the process of standardization. At present, 3GPP has close to 400 member companies from all over the world from network operators, base station, UE vendors, and chipset manufacturers to industrial and academic laboratories. During the standardization process, technical proposals are debated and discussed from different angles. A consensus-driven process decides which portions should be included in the next release of the standard. A large and complex system such as LTE has multiple functional components – from digital signal processing to core network optimization – and it is not possible for everything to be discussed in one group. Accordingly, 3GPP has a well-structured organization. As shown in Figure 1.4, 3GPP comprises four major working groups each in charge of a specific area. Each working group is further divided into subgroups for more specialized focus.

An understanding of standardization provides an insight into the features and timeline of future cellular systems. In the following chapters, the development of heterogeneous networks within the LTE-Advanced standardization framework will be presented in detail. The focus of this book will be on RAN working group 1 which pioneers the physical layer signal processing and related standardization aspects. Wherever applicable, issues related to the other Radio Access Network (RAN) working groups will also be presented.

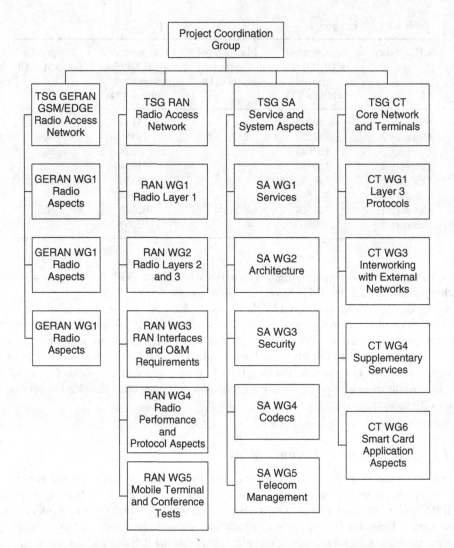

Figure 1.4   3GPP organizational chart showing various working groups

References

[1] Strategy Analytics (2013) Handset data traffic (2001–2017), June 2013. Strategy Analytics.
[2] Cisco (2013) Cisco visual networking index: Global mobile data traffic forecast update, 2012–2017, February 2013. Cisco.
[3] Chunfeng, C. (2013) LTE Evolution, Rel-12 and Beyond. *UK China Science Bridges: R&D on (B)4G Wireless Mobile Communications Workshop*, May 2012.
[4] Haszeldine, D. (2012) The economics of heterogeneous networks. *HetNet - The Base Station Conference*, September 2012.
[5] Mallinson, K. (2012) The 2020 vision for LTE. Available at http://www.3gpp.org/2020-vision-for -LTE (accessed November 2013).

[6] Pike, A. (2012) The impact of small cell deployment on radio network energy consumption. *HetNet - The Base Station Conference*, September 2012.

[7] DAS Forum (2013) Distributed antenna systems (DAS) and small cell technologies distinguished. Available at http://www.thedasforum.org/resources/das-and-small-cell-technologies-distinguished/ (accessed November 2013).

[8] Sesia, S., Toufik, I., and Baker, M. (2011) *LTE - The UMTS Long Term Evolution: From Theory to Practice*. United Kingdom: John Wiley & Sons Ltd.

Part One

Overview

2

Fundamentals of LTE

2.1 Introduction

LTE has been designed to improve the overall quality of experience of the end users relative to previous cellular systems. The number of users supported by LTE is higher and the range of applications provided is broader. All these are made possible by extensive design and optimization of almost all layers of the cellular system [1–3]. In this chapter we cover the technologies that have been incorporated within the LTE framework. We discuss the advances in the Core Network and the physical layer technologies that form the backbone of LTE.

The focus of the chapter is on the fundamentals of physical layer signal processing incorporated in LTE such as orthogonal frequency division multiplexing (OFDM) and multiple-input-multiple-output (MIMO) antenna communications. The theoretical fundamentals of these technologies are described as they will be needed in later chapters where we demonstrate how they are practically implemented in LTE.

The overall network architecture of LTE is shown in Figure 2.1. The LTE network consists of just two layers: the Evolved Universal Terrestrial Radio Access Network (E-UTRAN) which is the radio access network to the UE and the Evolved Packet Core (EPC) which is the Core Network. The architecture of the core network is also referred to as the Service Architecture Evolution (SAE) and the combination of E-UTRAN and EPC/SAE is also called the Evolved Packet System (EPS). Subsequently, we use the terms EPC and Core Network (CN) interchangeably. Likewise, in subsequent chapters the term LTE will be used to denote the LTE E-UTRAN or, more specifically, the PHY (Physical Layer) and Medium Access Control (MAC) layers of the E-UTRAN.

1. **E-UTRAN:** The E-UTRAN is the first point of entry for a UE to the LTE network. The E-UTRAN protocols cover the communication process between the UE and the network over the wireless link. E-UTRAN is responsible for the transmission/reception of radio signals to and from a given UE and the associated digital signal processing. E-UTRAN also includes the medium access control mechanisms by which multiple UEs share the common wireless channel. Other functions of E-UTRAN are to ensure link level reliability, segmentation, and reassembly of higher-layer Protocol Data Units (PDUs) and IP header compression. These protocols between the UE and the E-UTRAN are collectively referred to as Access Stratum (AS) protocols. The enhanced Node B (eNodeB or eNB) is the single

Heterogeneous Networks in LTE-Advanced, First Edition. Joydeep Acharya, Long Gao and Sudhanshu Gaur.
© 2014 John Wiley & Sons, Ltd. Published 2014 by John Wiley & Sons, Ltd.
Companion Wesite: www.ltehetnet.com.

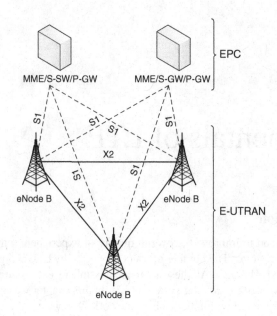

Figure 2.1 Overall LTE Architecture

logical node in the E-UTRAN. The job of the eNodeB is to implement the AS protocols. In Section 2.3, we discuss the eNodeB and E-UTRAN in more detail.

2. **EPC:** The UE communicates with the EPC through the E-UTRAN. When a UE powers on, the EPC is responsible for authentication and the initial connection establishment needed for all subsequent communication. Unlike Universal Mobile Telecommunications Systems (UMTS), LTE has an all-IP architecture that supports only packet-switched (PS) data. This leads to easier integration with the internet, thus enhancing user experience. The EPC is responsible for allocating IP addresses to the UE and forwarding/storing packet data to and from the UE to the external IP network. The signaling and protocols between the UE and the EPC are collectively called the Non Access Stratum (NAS) to distinguish it from the Access Stratum protocols. The EPC layer comprises several logical nodes such as Mobility Management Entity (MME) and Serving and Public Data Network (PDN) Gateways (S-GW and P-GW). These are discussed in more detail in Section 2.2.

The flat E-UTRAN- and EPC-based architecture of LTE is a simplification and improvement of previous hierarchical network architectures such as UMTS [4]. To understand this, consider the UMTS network architecture depicted in Figure 2.2. In UMTS, the core network is divided into multiple hierarchical entities. In the access side, several Base Transceiver Stations (BTS) are controlled by one Base Station Controller (BSC)[1] which performs the medium access control, reliability, segmentation, and reassembly and other functions. If a BSC malfunctions, a large coverage area incorporating many BTSs is affected. In LTE, the functionalities of the

---

[1] The term BSC is used primarily for GSM networks. For 3G networks the corresponding term is RNC (Radio Network Controller).

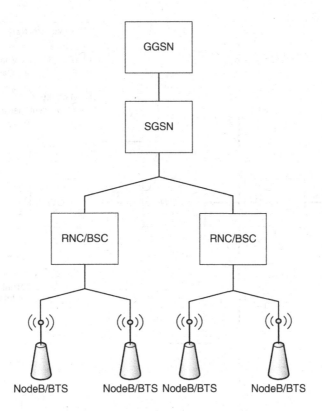

Figure 2.2 Overall UMTS Architecture

UMTS BTS and BSC are combined in the eNodeB and hence the problem does not arise. The flat, all-IP network architecture of LTE makes it efficient, scalable, and easy to integrate with existing data networks.

2.2 LTE Core Network

Recall that LTE is based on packet mode communications. However, it also operates in a *connection-oriented* mode which means that the entities at the end points of a communication processed first have to establish an end-to-end logical connection before any data can be transmitted. Note that this is different from circuit switched communication such as the public switched telephone network (PSTN) where there is an actual *physical* connection between the entities. For LTE, the two entities between which a logical connection has to be established prior to actual data transfer are the UE and the PDN Gateway (which will be explained later in this section). This logical connection is called the *EPS Bearer*. In other words, an EPS bearer is an end-to-end IP packet flow with a defined QoS. Many QoS parameters are defined in LTE such as scheduling priority, throughput, and reliability of transmission. A UE can be associated with a maximum of 8 EPS bearers. The differences in these bearers are as follows.

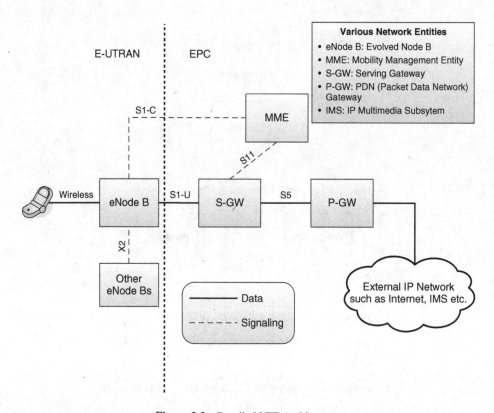

Figure 2.3 Detailed LTE Architecture

- Different bearers are associated with different PDN gateways.
- Different bearers are associated with different applications (IMS- based voice/internet data) and thus have different QoS requirements.
- One bearer is the *default bearer* which is set up initially by the network before the UE has started any application, and the other bearers are *dedicated bearers* set up for specific applications.

Management of EPS bearers and data transfer between the UE and external packet data network falls under the purview of the CN. The architecture of the CN is depicted in Figure 2.3. The CN has a control and a user plane, described in the following sections.

2.2.1 *Control Plane*

The main entity in the control plane is the Mobility Management Entity (MME). The MME is responsible for NAS signaling between the UE and the CN. NAS signaling takes place when a UE initially powers on and then attaches itself to the LTE network. The MME is responsible for the establishment, maintenance, and release of the EPS bearers. The MME also keeps track of the UEs when they are in idle mode (idle and connected modes are discussed in Section 2.3.1). The MME also manages interconnection between LTE and other 2G-/3G-based cellular networks.

2.2.2   User Plane

There are two main logical entities in the user plane: P-GW and S-GW.

1. **Packet Gateway (P-GW):** The P-GW is a node that connects a UE to an external packet data network (PDN) such as the internet and IMS (IP Multimedia Subsystem). It acts as the default router for the UE and is responsible for IP address allocation for the UE. A UE can have multiple IP addresses corresponding to connections with different P-GWs (different PDNs). The P-GW can also perform flow control and QoS enforcement of the IP packet flow to the UE.
2. **Serving Gateway (S-GW):** The S-GW is responsible for overall packet routing and forwarding to and from the UE. In case of inter-eNodeB handovers, the S-GW acts as the local mobility anchor. For example, if the UE is in idle mode and data comes for the UE from the external network, the S-GW buffers the data packets and requests the MME to page the UE.

Note that an EPS bearer is a data connection from the UE to the eNodeB to the S-GW and finally to the P-GW. It is managed by the MME but does not contain it, as no data transfer takes place through the MME.

2.2.3   Practical Implementations of the Core Network

The reader may have noted that we have been referring to the MME, S-GW, and P-GW as *logical entities*. This is because they refer to specific functionalities of the CN. In an actual core network implementation, some of these functionalities may be combined in a common hardware unit. Some of the possible options are as follows.

1. MME and S-GW combined as one unit, while the P-GW is in a separate unit. Legacy 3G CNs have a similar design and so this option allows for easier interoperability between a LTE and a 3G network.
2. MME is one unit and S-GW and P-GW are combined in another unit. This combines the two logical nodes in the user plane into one common physical unit. Since both units are essentially IP packet routers with evolved functionalities, this could be advantageous from an implementation viewpoint.
3. All three logical entities are separate physical entities. Since the areas covered by a MME, S-GW, and P-GW are different, this architecture provides the most flexible design.

Integrating more logical entities in a physical node will normally reduce latency and delay as the internode signaling is reduced; however, complexity of an individual physical node increases. Further, in the event of a node failure more CN functionalities are affected. Usually the choice of a specific physical implementation will depend upon the equipment manufactured by EPC vendors and also the strategy of the network operator.

The process by which different logical entities are assigned to a UE once it switches on is summarized in the following.

1. The UE selects the eNodeB based on received signal strength measurements. This is explained in detail in Section 2.3.1.
2. The eNodeB selects the MME based on load-balancing algorithms that ensure that the CN signaling load in a MME does not become too high. The eNodeB can also choose the previous MME that the UE was connected to.
3. The MME selects the S-GW based on geographical location of the UE. It picks up a S-GW that serves the eNodeB to which the UE is associated.[2] Load-balancing algorithms are used to choose between multiple S-GWs.
4. The UE chooses a PDN such as the internet based on the application it wishes to run.
5. The MME chooses a P-GW for the UE based on the PDN.

To optimize the performance of heterogeneous networks, additional functionalities in the CN may be needed. For example, due to the access to femtocells of the closed subscriber group (CSG), a CSG server is needed in the CN to manage femto-based communications. This is discussed further in Chapter 5.

2.3   LTE Radio Access Network

Recall that E-UTRAN is responsible for radio access of the UE to the LTE network. The E-UTRAN functionalities are implemented by a single node, the eNodeB. The eNodeB has several important functionalities, most of which are related to the wireless channel and its properties. Signals transmitted over the wireless channel suffer attenuations in amplitude and phase. The E-UTRAN therefore performs sophisticated Physical Layer (PHY) signal processing to ensure a reliable reception. The wireless channel is of limited bandwidth and is a broadcast medium. When multiple UEs access the channel simultaneously, they interfere with each other; the situation is particularly true for heterogeneous networks. The E-UTRAN therefore performs Radio Resource Management (RRM) and Medium Access Control (MAC) functionalities to allocate the channel resources among the various users in a fair manner. Note that the errors introduced by the wireless channel cannot be totally corrected by the PHY and MAC/RRM functionalities and hence an Automatic Repeat Request (ARQ) based flow control is required between the eNodeB and the UE. This functionality is performed by the Radio Link Control (RLC) layer. The Packet Data Convergence Protocol (PDCP) compresses the IP headers from higher layers so that they are suitable for transmission over the wireless channel. Finally, the Radio Resource Control (RRC) layer handles the control plane functionalities. These are all described in more detail in the following sections.

2.3.1   Control Plane

The LTE E-UTRAN control plane is shown in Figure 2.4. In this section we discuss the RRC layer; the remaining layers are discussed in the following section.

The RRC layer is broadly responsible for all AS control signaling, i.e. control information needed for the UE to access the E-UTRAN via the wireless channel. At any given time, a

---

[2] In LTE, a group of adjacent MMEs in a geographical region form a Tracking Area (TA). A group of TAs are served by a pool of S-GWs called the S-GW Serving Area. The MME will thus pick up a S-GW from the S-GW Serving Area that includes the TA of the UE.

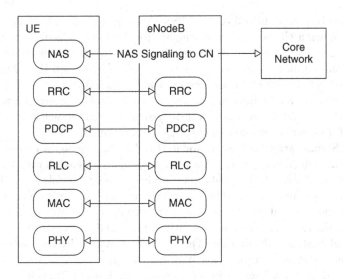

Figure 2.4 E-UTRAN Control Plane Protocols

UE is in either of the following two RRC states: RRC_IDLE and RRC_CONNECTED. The AS control signaling by the RRC depends on the UE state. Before going into the details of these two states, their motivation should be stated. The UE is more closely controlled and capable of complete data transmission and reception in the RRC_CONNECTED state. However, this increases the number of tasks that have to be performed by the UE and reduces the battery life. The UE therefore periodically switches to the RRC_IDLE state, where it is in partial sleep mode. It forgoes many tasks that it would have otherwise performed in the RRC_CONNECTED state, prolonging the UE battery life. A UE in RRC_IDLE mode performs only the most important control signaling but, to be able to transmit and receive data, it has to be in the RRC_CONNECTED state.

In addition to the RRC control signaling, the E-UTRAN also carries NAS signaling messages between the UE and the CN.

2.3.1.1 RRC IDLE State

When a UE powers on, it is initially in the RRC_IDLE state as a RRC connection has not yet been established. The UE has to first decide which eNodeB to associate with for subsequent control and data communications. This is called *cell selection* or deciding a cell to *camp on*, and involves the following processes.

1. **Initial Cell Detection:** LTE is an asynchronous network. This means that different eNodeBs transmit with their own symbol and frame timings. In order for any meaningful communication with the eNodeB, a UE has to first synchronize with it. In LTE each eNodeB has a unique cell ID which takes a value from 0 to 503. Each eNodeB periodically transmits two signals called the Primary Synchronization Signal (PSS) and Secondary Synchronization Signal (SSS). The PSS/SSS from different eNodeBs are differentiated

by their cell IDs. A list of cell IDs and PSS/SSS sequences for all eNodeBs are stored in each UE. When a UE powers on, it searches for PSS/SSS sequences in the radio signal that it receives by correlating the received signal with the known PSS/SSS sequences. The correlation process therefore not only gives the UE the cell IDs of the eNodeBs in the vicinity, but also their symbol timings. The UE then decodes the Physical Broadcast Channel (PBCH) whose information content is referred to as the Master Information Block (MIB). The MIB contains important system information such as system bandwidth, number of transmitting antennae, and system frame number.

2. **Received Signal Strength Measurement:** The UE then measures certain signals called Reference Signals (RS) from each eNodeB to determine the eNodeB with the best downlink gains. Also recall that a LTE signal can be centered around multiple carrier frequencies. A UE can be programmed *a priori* by the service provider so that there are some preferred carriers defined. The UE maintains a list of the eNodeBs with strongest link gains for each allowed carrier in which it can subsequently transmit and receive data.

3. **Reading of System Information:** The UE now reads the System Information Blocks (SIBs) which contain control signal from both AS (generated at RRC level) and NAS (generated by the CN). These SIBs are carried in the Physical Downlink Shared Channel (PDSCH). Specifically, the UE decodes SIB-1 which provides information about the public land mobile network (PLMN) identity of the eNodeB. The UE checks if it is permitted to associate with a eNodeB with the decoded PLMN identity. If it is not permitted (e.g. the UE might have been trying to camp on to an eNodeB that belongs to a different operator), it goes back to the list of eNodeBs and chooses the eNodeB with the next strongest link. It then repeats step (3).

4. **Final Association:** Once a suitable eNodeB has been found from step (3), the UE can further check if the received signal strength is over a certain threshold in order to finalize the association.

Note that the various physical channels (PSS, SSS, PBCH, PDSCH) and reference signals are discussed in more detail in Chapters 3 and 4.

Once the UE has camped on a cell, it continues to monitor the other neighboring eNodeBs. A RRC_IDLE UE can undergo cell reselection if it detects a eNodeB with stronger link gain. Note that the radio link quality is the most important criterion for cell selection and reselection. The radio link is however time varying; if this was the only criterion, there could be a ping-pong effect where the UE switches connection between a group of eNodeBs continuously. This is especially true for heterogeneous networks where there may be a dense deployment of nodes. There are therefore other criterion such as eNodeB priorities, especially for cell reselection. The threshold mentioned in step (4) could be adjusted once a UE performs initial cell selection, in order to prevent it from frequent reselection.

A UE in RRC_IDLE mode monitors the paging channel periodically to see if there are any incoming calls for it; if so, it switches to the RRC_CONNECTED mode.

2.3.1.2   RRC CONNECTED State

In the RRC_CONNECTED state, the UE can receive and transmit data from the E-UTRAN via the PDSCH and the corresponding uplink channel, the Physical Uplink Shared Channel

(PUSCH), respectively. To enable this, the UE monitors an associated control channel, the Physical Downlink Control Channel (PDCCH), from the eNodeB. The UE provides channel quality and feedback information to the eNodeB while in RRC_CONNECTED state.

In the connected mode, the UE can change cell associations as it did in idle mode. In the connected mode however, this is initiated and managed by the network and is called *handover*. Similar to cell reselection, handover is based primarily on link qualities but includes other parameters.

2.3.2   User Plane

The protocols in the E-UTRAN user plane are shown in Figure 2.5 and are described in the following sections.

2.3.2.1   Physical (PHY) Layer

The PHY layer is responsible for all radio signal transmission and reception between the UE and the eNodeB and the associated digital signal processing. This includes orthogonal frequency division multiplexing (OFDM) for most efficient spectrum usage, adaptive modulation and channel coding (AMC) to adapt to the variable channel conditions, and transmission/reception from multiple antennae for exploiting diversity and spatial multiplexing gains. The receiver (UE for downlink and eNodeB for uplink) employs advanced receiver processing filters to minimize error performance. Details of the LTE PHY framework are provided in Chapter 4.

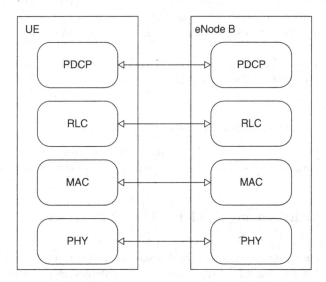

Figure 2.5   E-UTRAN User Plane Protocols

2.3.2.2   Medium Access Control (MAC) Layer

This layer performs multiplexing of different types of data from the upper layer (e.g. RRC control messages or application-specific data) into MAC PDUs that are passed onto the PHY layer for transmission. The MAC performs dynamic scheduling to decide which UEs should be eligible for transmission/reception at a given time and over a given frequency band. The MAC scheduler ensures that the resource allocation to the UEs is fair in time and frequency. The scheduler design is crucial for ensuring the overall performance of the network. The gains of a sophisticated PHY scheme such as advanced MIMO processing can be greatly boosted by an intelligent scheduler design, and the reverse is also true. The optimum scheduler for a homogeneous network with sparse macrocells and low UE load can be very different from that of a dense heterogeneous network.

The MAC also performs Hybrid ARQ (HARQ) to correct the errors in the link level communication between the UE and the eNodeB. LTE allows for simultaneous transmission of 8 parallel stop-and-wait HARQ processes to improve the overall throughput. The maximum number of retransmissions in a HARQ process is 8.

2.3.2.3   Radio Link Control (RLC) Layer

The HARQ operation in the MAC layer may not be able to correct all the errors in the link layer. This is handled by an ARQ process in the RLC layer. For example, it is possible that the maximum of 8 retransmissions has been reached in the MAC HARQ and yet the MAC PDU have not been received successfully. The MAC informs the RLC, and the RLC can recode/re-segment the packet for a new transmission.

Alternatively, the HARQ receiver in the MAC may have transmitted a NACK (negative acknowledgement) for a received packet. It would then expect the transmitter to resend the packet shortly. However the NACK message can itself become corrupted by the wireless channel and the transmitter can interpret it as an ACK. The transmitter would then transmit a new packet. When the receiver sees a fresh packet, it informs the RLC which takes care of the situation.

The RLC layer also performs segmentation and reassembly of higher-layer packets to adapt to the time-varying nature of the wireless channel fading.

2.3.2.4   Packet Data Convergence Protocol (PDCP) Layer

The PDCP layer performs IP header compression as the standard IP headers can be too large for transmission over the wireless channel. It also takes care of certain security aspects.

2.4   Connectivity Among eNodeBs: The X2 Interface

As seen in Figure 2.1, the various nodes in the LTE network are interconnected to allow them to communicate and share information. In this section, we focus on the X2 interface which connects two eNodeBs. The X2 interface allows two eNodeBs such as a macrocell and a picocell to 'talk' to each other and thus implement many of the advanced interference coordination algorithms that will be described in the subsequent chapters.

Table 2.1 Latency and throughput of some commonly used mediums for backhaul

Medium	Latency (one way, round trip, ms)	Throughput (average, Mbps)
Fiber	5–10	100–1000
Cable	25–35	10–100
DSL	15–60	10–100
Wireless	5–35	10–100

The standards describe X2 as a point-to-point logical interface between two eNodeBs. This logical interface is implemented over an actual physical connection between the two eNodeBs. In practical implementations, the medium of physical connection is vendor-specific and can be fiber, microwave, or ethernet. It is also possible that there is no direct physical connectivity between the eNodeBs, but the X2 interface algorithms can still be realized as the eNodeBs are connected via the EPC (e.g. two eNodeBs may be served by the same MME). Different mediums will lead to different latency and throughput capabilities as shown in Table 2.1 [5, Table 6.1-1]. The choice of medium will determine the performance of algorithms that are subsequently implemented over X2. For example, a dynamic interference coordination algorithm that requires fast exchange of interference information between eNodeBs may not be possible over a medium with high latency.

A LTE eNodeB can identify and authenticate a neighboring eNodeB. This can be configured manually by the operator who set up the network. Since this process is cumbersome and unreliable, especially when new eNodeBs are dynamically deployed in the network, LTE also provides for a function called the Automatic Neighbor Relation Function (ANRF) which performs smart neighbor discovery. All eNodeBs broadcast information such as cell identities that a UE in their coverage area can measure during the cell association process. After association, an eNodeB can ask its UEs to report these measurements and use them for neighbor discovery. ANRF is an example of a broader class of LTE functionalities called SON (Self-organized Network) which were introduced for automatic self-configuration and operation of the LTE network.

Once the neighboring cells have been identified, a eNodeB triggers the X2 setup procedure. This involves the initiating eNodeB sending a X2_SETUP_REQUEST message and the receiving eNodeB replying with an acknowledgement.

The X2 protocol stack is given in Figure 2.6. It has a control plane for exchanging signaling messages between the eNodeBs and a user plane for exchanging PDUs. To understand the basic nature of these two planes, consider the example of a UE handover from a source eNodeB to a target eNodeB. The control plane first carries signaling messages that are required to set up the handover procedure. The user plane then transfers the user data from the source eNodeB to the target eNodeB. The two protocol stacks thus vary slightly depending on the nature of the messages they carry. For example, the Control Plane has Stream Control Transmission Protocol (SCTP) as the transport protocol to ensure higher reliability of the control messages while the user plane uses the less reliable but faster User Datagram Protocol (UDP). X2-based handover is very important for macro–pico heterogeneous networks and is covered in greater detail in Chapter 5.

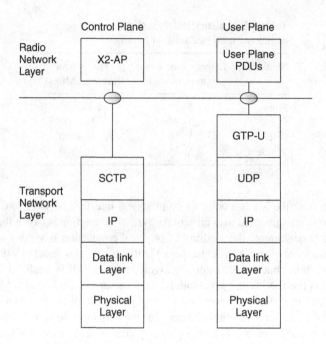

Figure 2.6 Protocols for X2 signaling bearer and data stream transmission

Broadly, two types of information may be exchanged over X2. These are discussed in the following sections.

2.4.1 Load- and Interference-Related Information

The aim of load balancing is to distribute UE traffic load uniformly among a group of eNodeBs. In UMTS, the RNC took care of load management between the base stations that were connected to it. In LTE, the eNodeBs themselves exchange this information via the X2 interface. The eNodeBs can exchange usage pattern of their bandwidth (for example what percentage of their allocated bandwidth is being used by real-time vs non-real-time traffic) to convey load-related information. Based on information exchange over X2, the eNodeBs jointly perform load management by optimizing cell reselection and handover parameters for current UEs that are already associated, or new UEs who may request service in the future.

Separate indication messages are exchanged between neighboring eNodeBs to denote the real-time interference being faced by an eNodeB from a neighboring eNodeB. An eNodeB can also indicate to a neighboring eNodeB if it is planning to increase traffic over a certain frequency band which would cause interference to its neighbor. Such messages are available for both downlink and uplink interference scenarios and are covered in more detail in Chapter 4.

2.4.2 Handover-Related Information

UE handover can also take place directly between two eNodeBs by signaling over X2. If an X2 interface is not present, handover between two eNodeBs takes place via the CN. For X2-based

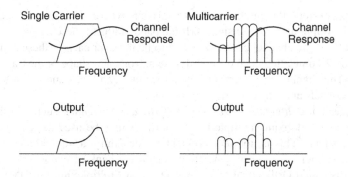

Figure 2.7 Single-carrier versus multicarrier transmission over frequency-selective fading channels

handover, the MME is informed only after successful completion of the handover. X2-based handover is initialized by the source eNodeB by sending a HO_REQUEST message to the target eNodeB. Upon receiving the request message, the target eNodeB allocates resources needed for the handover and then responds with a HO_REQUEST_ACK message. In the subsequent handover process, UE PDUs are transferred from the source to the target eNodeB along with other UE information via the User Plane of X2. User PDU transfer can either be *seamless*, which minimizes the time taken by the handover process, or *lossless*, which maximizes the reliability of the user data. The source eNodeB chooses the handover mode based on the Quality of Service (QoS) requirements of the associated EPS bearer.

2.5 Technologies in LTE

In this section we discuss some of the signal processing technologies in the LTE PHY layer. We present a brief recapitulation of the theoretical fundamentals, assuming that the reader has prior knowledge. We skip the details of mathematical modeling and focus more on insights gained from these technologies. We also introduce the issues related to their practical implementation, which are discussed in more detail in Parts II and III.

2.5.1 Orthogonal Frequency Division Multiplexing

OFDM is both a modulation and a multiple access scheme that is optimal for high-rate data transmissions over wideband channels. OFDM is an example of a multicarrier modulation scheme which is different from single-carrier modulation schemes as used in previous cellular systems such as Wideband Code Division Multiple Access (WCDMA) and Global System for Mobile communications (GSM).

To understand the importance of multicarrier systems consider Figure 2.7 which depicts the frequency domain representation when a signal is transmitted over a frequency-selective channel. For a single-carrier wideband transmission, different parts of the signal spectrum are attenuated differently. This is because in the time domain, there are multipaths which lead to inter-symbol interference (ISI). The output spectrum looks very different from the input and complex equalizers are needed to reverse the effects of the channel. Complex channel equalizers have to be designed for single-carrier cellular systems such as GSM.

In contrast, consider the multicarrier transmission where the same signal spectrum is composed of multiple narrowband signals. If the bandwidths of these narrowband signals are designed carefully, the channel can be assumed to be flat over each of these bands. Although different bands become attenuated differently, the narrowband shape of the transmit spectrum is preserved. The channel equalizers needed are simpler and hence more reliable. It is easier to decode the signals at each subcarrier.

The next question is *how do we generate a multicarrier signal*? There are many ways to do so. In the analog domain, multiple bandpass filters may be used to generate and receive a multicarrier signal. This is not very efficient for designing a flexible and robust system. One easy and efficient method of generation is to use a bank of exponentials in the digital domain. This is called Orthogonal Frequency Division Multiplexing (OFDM), explained in the following section.

2.5.1.1   Basics of OFDM Transmission

Consider a wireless channel with impulse response $h(n)$ and let the input signal be $x(n)$. The received signal $y(n)$ is given by

$$y(n) = h(n) \otimes x(n) + z(n), \tag{2.1}$$

where $z(n)$ is the receiver noise. Let us first consider a narrowband signal centered around frequency w_0 for which $x(n) = e^{jw_0 n}$ for $-\infty < n < \infty$. In this case the output signal is given by $y(n) = H(w_0)e^{jw_0 n} + z(n)$, where $H(w)$ is the spectrum of $h(n)$. We observe that $y(n)$ is the same exponential $x(n)$ with an amplitude and phase change, given by $H(w_0)$ plus noise. This is because exponentials are eigenfunctions of linear systems. This leads to the idea of generating a wideband multicarrier system: if an input data vector $s(n) = [s(0), \cdots, s(N-1)]$ is to be transmitted, then modulate it by a bank of exponentials. In this case the transmit signal becomes

$$x(n) = \sum_{k=0}^{N-1} s(k)e^{jw_k n}, \; -\infty < n < \infty. \tag{2.2}$$

In the frequency domain, the transmission can be represented by

$$Y(w_k) = H(w_k)s(k) + Z(w_k), \; 0 \leq k \leq N - 1. \tag{2.3}$$

The whole system is therefore equivalent to a set of N parallel channels. Data symbol $s(k)$ is transmitted in channel k with no interference from symbols in other channels; ISI is therefore avoided.

However the signal in Equation (2.2) is infinite and cannot be an OFDM signal. A finite duration OFDM signal is designed by truncating Equation (2.2) to a finite number of bits. We now show that this can be done efficiently using the Discrete Fourier Transform (DFT).

Let $x(n)$ be a N-point sequence with N-point DFT $X(k)$. Recall that these are related by the DFT and inverse DFT (IDFT) operations respectively, defined as

$$X(k) = \sum_{n=0}^{N-1} x(n)e^{-j\frac{2\pi n}{N}k}, \; 0 \leq k \leq N - 1 \tag{2.4}$$

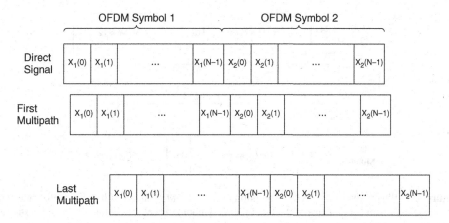

Figure 2.8 Reception of two OFDM symbols in the presence of ISI (to simplify the illustration, the multipath components are shown with only their time delays and not with their amplitude attenuations)

$$x(n) = \sum_{k=0}^{N-1} X(k) e^{j\frac{2\pi k}{N}n}, \ 0 \le n \le N-1. \tag{2.5}$$

Compare Equations (2.2) and (2.5) and note their similarity. By choosing $X(k) = s(k)$ and $w_k = 2\pi k/N$, the two expressions are the same for sample instants $0 \le n \le N-1$. An OFDM symbol can therefore be generated via DFT modulation of an input symbol vector $[s(0), \cdots, s(N-1)]$ as

$$x(n) = \sum_{k=0}^{N-1} s(k) e^{j\frac{2\pi k}{N}n}, \ 0 \le n \le N-1. \tag{2.6}$$

Let us introduce some terminology. The vector $x(n) = [x(0), \cdots, x(N-1)]$ is referred to as an OFDM symbol from this point onwards. This is different from the input data symbols $s(0), \cdots, s(N-1)$. The samples $x(0), \cdots, x(N-1)$ are referred to as the samples of the OFDM symbol.

We have shown how to generate an OFDM symbol. However, in that process the infinite exponential sequences that were the eigenfunctions of the channel were truncated. Doesn't this contradict the property that multicarrier transmission is essentially transmission over a set of parallel channels?

For the purposes of illustration, consider two OFDM symbols each with N samples. Let the symbols be $x_1(n) = [x_1(0), \cdots, x_1(N-1)]$ and $x_2(n) = [x_2(0), \cdots, x_2(N-1)]$. Consider the serial transmission of these two symbols over a multipath channel $h(n)$, as shown in Figure 2.8. Observe that there are two kinds of ISI that occur at the receiver:

1. interference between samples belonging to two different OFDM symbols; and
2. interference between samples of the same OFDM symbol.

Because of these interferences, the parallel channel structure as given by Equation (2.3) cannot be achieved. Fortunately this structure can be restored by a technique called the *Cyclic Prefix* insertion. We show how this is achieved in the following section.

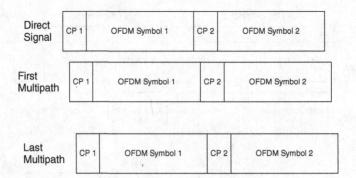

Figure 2.9 Illustration of the need of reception of two OFDM symbols with cyclic prefix; interference due to the multipath components of the first OFDM symbol falls entirely within the cyclic prefix region of the second OFDM symbol

2.5.1.2 Cyclic Prefix

Assume that the channel has L taps given by $h(n) = [h(0), \cdots, h(L-1)]$. Cyclic prefix insertion takes the last $L_p \geq L$ samples of the OFDM symbol and appends them at the beginning of the symbol.

The prefix insertion and subsequent symbol reception are depicted in Figure 2.9. If $L_p > L$, which in continuous time means that the length of the prefix is chosen to be longer than the maximum possible delay spread of the multipath channel, then the ISI between samples belonging to two different OFDM symbols are confined to the region of the prefix. If the received samples of the prefix are discarded at the receiver, then the inter-OFDM symbol interference can therefore be avoided. Since the prefix samples carry no new information that is not already contained in the OFDM symbol, there is no loss of performance if they are discarded at the receiver.

The prefix therefore functions similarly to the insertion of a guard time for eliminating inter-OFDM symbol interference. Instead of transmitting the prefix, the transmitter could have inserted a guard time interval between the transmissions of successive OFDM symbol transmissions. Even for this case, the inter-OFDM symbol interference would have been avoided. So why was the prefix transmitted? The answer is that the prefix also addresses the second kind of ISI, the interference between samples of the same OFDM symbol. A guard band insertion does not solve this, as we demonstrate analytically as follows.

Consider an OFDM symbol $x(n) = [x(0), \cdots, x(N-1)]$ and channel $h(n) = [h(0), \cdots, h(L-1)]$. Then $x_{CP}(n)$, the OFDM symbol with prefix, is

$$x_{CP}(n) = \begin{cases} x(n+N), & -L \leq n \leq -1 \\ x(n), & 0 \leq n \leq N-1. \end{cases} \tag{2.7}$$

From the properties of convolution, the received signal $y_{CP}(n) = h(n) \otimes x_{CP}(n) + z(n)$ is defined for $-L \leq n \leq N+L-1$. At the receiver, the first L samples are discarded and the next N samples are taken to match the sampling rate and timing of input signal $x(n)$. We denote this sequence $y(n) = [y(0), \cdots, y(N-1)]$.

The N-point DFT of $y(n)$ is given by

$$Y(k) = \sum_{n=0}^{N-1} y(n)e^{-j\frac{2\pi n}{N}k} \tag{2.8}$$

$$= \sum_{n=0}^{N-1} (h(n) \otimes x_{CP}(n))e^{-j\frac{2\pi n}{N}k} + Z(k) \tag{2.9}$$

$$= \sum_{n=0}^{N-1}\sum_{l=1}^{L-1} h(l)x_{CP}(n-l)e^{-j\frac{2\pi n}{N}k} + Z(k) \tag{2.10}$$

$$= \sum_{l=1}^{L-1} h(l)\left(\sum_{n=0}^{N-1} x_{CP}(n-l)e^{-j\frac{2\pi n}{N}k}\right) + Z(k). \tag{2.11}$$

If we denote the inner summation S and rewrite the equations with the substitution $m = n - l$, we have

$$S = \sum_{m=-l}^{N-1-l} x_{CP}(m)e^{-j\frac{2\pi(m+l)}{N}k} \tag{2.12}$$

$$= e^{-j\frac{2\pi l}{N}k}\left(\sum_{m=-l}^{-1} x_{CP}(m)e^{-j\frac{2\pi m}{N}k} + \sum_{m=0}^{N-1-l} x_{CP}(m)e^{-j\frac{2\pi m}{N}k}\right) \tag{2.13}$$

$$= e^{-j\frac{2\pi l}{N}k}(S_1 + S_2). \tag{2.14}$$

In S_1, set $x_{CP}(m) = x(m + N)$ from Equation (2.7). We next substitute $\acute{m} = m + N$ to obtain

$$S_1 = \sum_{\acute{m}=N-l}^{N-1} x(\acute{m})e^{-j\frac{2\pi(\acute{m}-N)}{N}k} \tag{2.15}$$

$$= \sum_{m=N-l}^{N-1} x(m)e^{-j\frac{2\pi m}{N}k}, \tag{2.16}$$

as $\acute{m}$ is a dummy index. Now observe that in S_2, $x_{CP}(m) = x(m)$. Substituting for S_1 and S_2 in (2.14) we obtain,

$$S = e^{-j\frac{2\pi l}{N}k}\left(\sum_{m=N-l}^{N-1} x(m)e^{-j\frac{2\pi m}{N}k} + \sum_{m=0}^{N-1-l} x(m)e^{-j\frac{2\pi m}{N}k}\right) \tag{2.17}$$

$$= e^{-j\frac{2\pi l}{N}k}\sum_{m=0}^{N-1} x(m)e^{-j\frac{2\pi m}{N}k} \tag{2.18}$$

$$= e^{-j\frac{2\pi l}{N}k}X(k), \tag{2.19}$$

where $X(k)$ is the N point DFT of $x(n)$. By substituting for S in Equation (2.11) we obtain

$$Y(k) = H(k)X(k) + Z(k), \ 0 \le k \le N - 1. \tag{2.20}$$

The set of N subcarriers therefore form a system with N parallel channels in the frequency domain with no interference. One interpretation is that the information symbols $s(k) = X(k)$ directly modulate the subcarrier k in the frequency domain.

Our analysis has demonstrated the need for cyclic prefix insertion. There are other equivalent methods of explaining why a prefix insertion ensures that the DFT of the output sequence is the product of the DFTs of the channel impulse response and the input signal vector:

1. the cyclic prefix converts the linear convolution between the input signal and channel to a circular convolution [6]; and
2. insertion of a cyclic prefix makes the finite time exponentials *appear* as periodic and infinite to the channel.

Finally, we note that if a guard band was transmitted instead of the cyclic prefix, S_1 in Equation (2.14) is zero and hence the parallel channel structure would not be realizable.

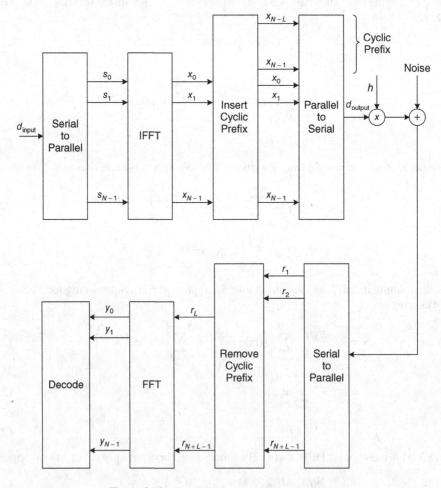

Figure 2.10   An OFDM transmitter and receiver

Since the prefix samples carry no new information, their transmission leads to an overhead in the transmission process. This is the trade-off in obtaining ISI-free transmission.

The OFDM transmitter and receiver structure are depicted in Figure 2.10 with the Fast Fourier Transform (FFT) block at the receiver and the inverse FFT (IFFT) block at the transmitter. Let us denote the symbol time of the input d_{input} as T_{input}. After serial to parallel operation, the rate of each of the output samples $s(0), \cdots, s(N-1)$ is reduced by a factor of N where N is the FFT order. The rest of the OFDM baseband operates at this rate. This is also illustrated graphically in Figure 2.11. This is another way of observing that the input d_{input} is wideband (high rate) and the N OFDM samples $x(0), \cdots, x(N-1)$ which are derived from $s(0), \cdots, s(N-1)$ are narrowband (lower rate).

2.5.1.3 Peak to Average Power Ratio (PAPR)

In continuous time, the OFDM signal $x(t)$ is obtained from the IDFT operation which sums up a large number of symbols, weighted by complex exponentials. The symbols come from Quadrature Amplitude Modulation (QAM)constellations and are independent and identically distributed (i.i.d.). From the central limit theorem, $x(t)$ approaches a Gaussian waveform. This means that $x(t)$ can have large amplitude variations. Power amplifiers that operate at the RF units can however only operate over a small dynamic range. Signal $x(t)$ with a large amplitude is therefore clipped, leading to non-linear distortions in the signal.

The peak to average power ratio (PAPR) of a signal $x(t)$ over a period T is a commonly used metric to measure the dynamic range of the input and hence the effect of non-linear distortions [2], defined as

$$\text{PAPR} = \frac{\max\limits_{0 \leq t \leq T} |x(t)|^2}{1/T \int_0^T |x(t)|^2 dt}. \tag{2.21}$$

A signal should have low PAPR to avoid non-linear distortions. However, OFDM signals have high PAPR. In the LTE downlink, Orthogonal Frequency Division Multiple Access (OFDMA) may therefore introduce non-linear distortion. No techniques for reducing PAPR in OFDM have been standardized in LTE, but the eNodeB vendors can implement their own solutions. In the uplink a variant of OFDM, single-carrier FDMA (SC-FDMA), has been used because of its superior PAPR property which is discussed in the following section.

2.5.1.4 OFDM in LTE

We have discussed several parameters of OFDM such as FFT size and CP length. In a LTE transmission the values of these parameters are selected based on various theoretical and practical considerations. In this section, we explain these with a series of examples.

Example 2.5.1 *Consider OFDM transmission over a fading channel with delay spread 5 μs. Let the UE velocity be 350 kmph and the carrier frequency be 4 GHz. How should the OFDM symbol duration and subcarrier spacing values be selected?*

The CP should be greater than or equal to the delay spread, so a prudent choice would be ~5 μs. As a rule of thumb, the CP length should not be more than 10% of the total symbol length to minimize overheads. The symbol length T_{OFDM} should therefore be of the order 50 μs.

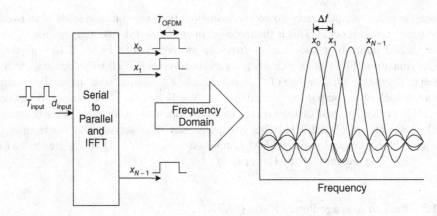

Figure 2.11 Another representation of the OFDM transmitter showing the relationship between input symbol duration, OFDM symbol time, and subcarrier spacing

Now consider the coherence time of the channel, T_C. From [7], this is related to the carrier frequency f_m and velocity v_m by

$$T_c = \sqrt{\frac{9}{16\pi} \frac{c}{v_m} \frac{1}{f_m}}. \tag{2.22}$$

For the values given in the example, $T_c = 0.3$ ms; $T_{OFDM} \sim 50\,\mu s$ is therefore also well within the coherence time of the channel and the signal waveform will experience correlated fading.

The bandwidth of the OFDM symbol should be inversely proportional to T_{OFDM}; it is therefore of the order 20 kHz. Since one symbol modulates one subcarrier in OFDM, the subcarrier spacing Δf should also be of the order 20 kHz so that the entire symbol can fit there. This is illustrated in Figure 2.11.

The channel delay spreads and coherence times in the above example are typical of the worst-case channel behavior. LTE has two kinds of prefix durations: normal and extended. The normal prefix duration is 5 μs for which T_{OFDM} and Δf are chosen to be 66.7 μs and 15 kHz respectively which leads to a prefix overhead of about 7.5%. An extended prefix of 17 μs can be used for channels with large delay spreads which affect the other parameters.

Example 2.5.2 *Consider two LTE systems with transmission bandwidths 1.4 MHz and 10 MHz. The subcarrier spacing is 15 kHz. What should the FFTs be in both systems?*

Refer to Figure 2.11 again. The input data stream has symbol duration T_{input} which is given by $T_{input} = T_{OFDM}/N$, where N is the FFT order. In terms of sampling rates, $R_{input} = NR_{OFDM}$ where R_{input} and R_{OFDM} are the respective sampling rates. It is now given that $R_{OFDM} = 15\,kHz$.

For signal bandwidth W, the sampling rate R_{input} should satisfy $R_{input} > W$ to prevent aliasing. In practice, R_{input} is chosen to be a multiple of 1.92 MHz which is half the chip rate of a UMTS system. This is done for backward compatibility.

Hence for $W = 1.4$ MHz the choice is $R_{input} = 1.92$ MHz. For $W = 10$ MHz the choice is $R_{input} = 1.92 \times 8 = 15.36$ MHz. The corresponding FFT sizes are 128 and 1024 respectively.

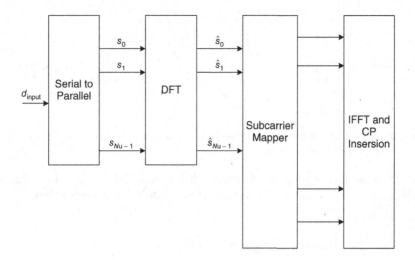

Figure 2.12   A SC-FDMA transmitter and receiver

Note that the FFT sizes turned to be exact exponents of 2, as needed for the radix-2 operation. This is no surprise as the value of $\Delta f = 15$ kHz had been selected to ensure this.

Example 2.5.3 *An LTE system with bandwidth 10 MHz is sampled at 15.36 MHz leading to a FFT size of 1024. How many of these subcarriers will be occupied with data?*

Recall that the subcarrier spacing is 15 kHz and the overall system bandwidth is 10 MHz. The number of occupied subcarriers is therefore 10 MHz/15 kHz = 667. The value chosen in LTE is 600 subcarriers which means that the signal actually spans a bandwidth of 9 MHz. The remaining subcarriers correspond to frequencies from 9 to 15.36 MHz, for which there are no data present.

LTE downlink employs QFDMA which enables multiple access among different UEs using OFDM transmission. In OFDMA, different UEs belonging to the same eNodeB are allocated a different number of subcarriers as per their individual traffic and QoS requirements. The channel gain experienced by a UE in a subcarrier is also taken into account during allocation. The multicarrier structure of OFDMA introduces the possibility of dynamic resource allocation [8].

As mentioned in Section 2.5.1, OFDM suffers from high PAPR. For this reason a new scheme called Single-Carrier FDMA (SC-FDMA) is used in the LTE uplink. This is shown in Figure 2.12.

The input signal to the IDFT is first *spread* in the frequency domain using another DFT block of smaller size, say N_u. The DFT size is equal to the actual bandwidth which is spanned by the UE signal. The IDFT size $N > N_u$ is equal to the system bandwidth. The subcarrier mapper chooses the N_u subcarriers allocated to the UE from the N subcarriers of the IDFT. Different UEs are allocated different sets of subcarriers.

This kind of DFT spreading ensures that a symbol is spread over multiple subcarriers. In contrast, during OFDM transmission one symbol is directly mapped to one (modulated) subcarrier. This effect of spreading over multiple subcarriers reduces the PAPR in SC-FDMA. The more apt name for SC-FDMA is therefore DFT-spread OFDM (DFT-S-OFDM).

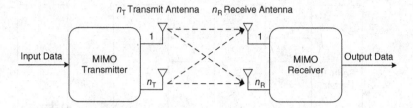

Figure 2.13   MIMO system showing the transmission from n_T transmit antennae to n_R receive antennae

Apart from the DFT spreading, the other transmission structure and the parameters for SC-FDMA in the LTE uplink are the same as those of OFDMA in the downlink.

2.5.2   *Multiple Antenna Communications*

MIMO improves performance of LTE systems by intelligent signal processing to exploit the presence of multiple spatial paths [9, 10]. A MIMO system is depicted in Figure 2.13.

The transmission between the n_T transmit antennae and the n_R receive antennae can be described by the baseband transmission model:

$$\mathbf{y} = \mathbf{Hx} + \mathbf{z} \qquad (2.23)$$

where $\mathbf{x}$ is the $n_T \times 1$ transmit signal vector, $\mathbf{H}$ is the $n_R \times n_T$ MIMO channel matrix, $\mathbf{z}$ is the $n_R \times 1$ vector of background noise plus interference and $\mathbf{y}$ is the $n_R \times 1$ received signal vector. When $n_T = 1$, the MIMO channel reduces to a Single-Input Multiple-Output (SIMO) channel, when $n_R = 1$ to a Multiple-Input Single-Output (MISO) channel and when $n_T = 1, n_R = 1$ then to the Single-Input Single-Output (SISO) channel.

For a MIMO-OFDM system such as LTE, the system can be modeled for each subcarrier k as

$$\mathbf{y}_k = \mathbf{H}_k\mathbf{x}_k + \mathbf{z}_k, \ 1 \leq k \leq N, \qquad (2.24)$$

where N is the total number of subcarriers used.

A MIMO channel provides multiple spatial paths between the transmitter and receiver that are not present for SISO. One advantage of multiple spatial paths is *spatial multiplexing* and another is *diversity*. In spatial multiplexing, multiple data streams can be simultaneously transmitted to the receiver where they are separately detected. Diversity, on the other hand, transmits a single data stream but increases the reliability of reception as compared to SISO. A MIMO system can use its antennae intelligently to trade off multiplexing with diversity. For a complete discussion on this topic the reader is referred to Zheng and Tse [11].

Both spatial multiplexing and diversity modes have been adopted in LTE. We now summarize their theoretical foundations.

2.5.2.1   **Spatial Multiplexing**

Using spatial multiplexing, multiple data streams can be simultaneously transmitted across the multiple spatial paths and decoded at the receiver. An independent transmitted data stream is

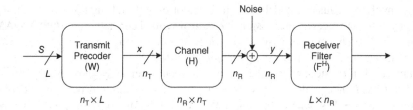

Figure 2.14 MIMO communication system with a linear transmitter and receiver

called a *layer* in LTE. The capacity of the MIMO channel is defined as

$$C_{\text{MIMO}} = \log \det(\mathbf{HQH}^H + \sigma^2\mathbf{I}), \tag{2.25}$$

where $\mathbf{Q} = E[\mathbf{xx}^H]$ is the covariance matrix of the transmitted signal vector $\mathbf{x}$ and σ^2 is the power of the receiver noise. Let P_{Total} be the total transmit power at the transmitter that has to be allocated across the n_T transmit antennae. C_{MIMO} then has to be maximized by optimizing $\mathbf{Q}$ under the transmit power constraint Trace(Q) $\leq P_{\text{Total}}$.

Let the singular value decomposition (SVD) of the channel matrix be $\mathbf{H} = \mathbf{U\Sigma V}^H$. The matrix $\mathbf{V}$ gives the *eigenmodes* or *eigendirections* of the channel correlation matrix $\mathbf{H}^H\mathbf{H}$. It can be shown that by choosing the transmit signal vector $\mathbf{x} = \mathbf{Vs}$ where $\mathbf{s}$ is the transmitted data stream, the MIMO channel transmission can be decomposed into a series of parallel interference-free channels $y_k = \lambda_k s_k + z_k$ whose capacities [12] are given by

$$C_k = \log\left(1 + \frac{\lambda_k^2 p_k}{\sigma^2}\right), \quad 1 \leq k \leq r \leq \min(n_R, n_T), \tag{2.26}$$

where λ_k^2 is the kth eigenmode of the channel correlation matrix and p_k is the power allocated in the kth eigenmode. The number of parallel channels is equal to the rank r of the channel matrix $\mathbf{H}$. The MIMO capacity is given by $C_{\text{MIMO}} = \sum C_k$ with the total power constraint $\sum p_k = P_{\text{Total}}$. It can be shown that the optimal power allocation is given by the waterfilling expression,

$$p_k = \left(\mu - \frac{\sigma^2}{\lambda^2}\right)^+, \tag{2.27}$$

where $a^+ = \min(a, 0)$ and μ is the waterfilling level. The capacity is achieved by choosing the transmit symbols s_k from a Gaussian codebook with power given by p_k. From Equation (2.26), it becomes clear that the MIMO capacity scales with $r \leq \min(n_R, n_T)$. In the absence of channel correlation, $r = \min(n_R, n_T)$.

In this analysis we assumed that the channel $\mathbf{H}$ was perfectly known at the transmitter and receiver. In LTE, the receiver measures the channel and feeds back quantized information about the channel which the transmitter obtains after a finite delay. Information-theoretic analysis of MIMO systems with no assumption of channel knowledge at the transmitter leads to notions of *ergodic capacity* for slow-fading channels and *outage capacity* for fast-fading channels. For a more detailed understanding of such issues the reader is referred to Tse and Viswanath [13].

Note that the information-theoretic sum capacity analysis answers the question *what are the theoretical limits of capacity gain via MIMO?* It also suggests a theoretical achievability strategy involving Gaussian codebooks. It does not directly tell us how to design a system with transmitter and receiver processing of practical symbol constellations (QPSK/QAM). The information-theoretic analysis however provides important insights which we will now use to derive practical transceiver processing methods for MIMO.

Assume a MIMO system with a linear transmitter and receiver, as shown in Figure 2.14. The input data $\mathbf{s}$ has L independent layers and a linear matrix operation is performed to obtain the transmit vector $\mathbf{x} = \mathbf{Ws}$, where the matrix $\mathbf{W}$ is called the *precoder* in LTE. At the receiver, a filter $\mathbf{F}^H$ is applied to separate the layers and the filter output is then decoded. The overall received signal is given by

$$\mathbf{r} = \mathbf{F}^H\mathbf{y} = \mathbf{F}^H\mathbf{HWs} + \mathbf{F}^H\mathbf{z}. \tag{2.28}$$

It is instructive to look at the transmission and reception of a single layer k. The vector $\mathbf{w}_k$ which is the kth column of $\mathbf{W}$ is the *precoding vector* for the kth stream. The corresponding received filter is $\mathbf{f}_k^H$. The corresponding received signal is

$$\mathbf{r}_k = \mathbf{f}_k^H\mathbf{Hw}_k s_k + \mathbf{f}_k^H\mathbf{n}_k, \tag{2.29}$$

where $\mathbf{n}_k = \sum_{j \neq k}\mathbf{Hw}_j s_j + \mathbf{z}$ constitutes of inter stream layer interference and noise. The covariance of $\mathbf{n}_k$ is

$$\mathbf{R}_k = E[\mathbf{z}_k\mathbf{z}_k^H] = \sum_{j \neq k}\mathbf{Hw}_j\mathbf{w}_j^H\mathbf{H}^H + \sigma^2\mathbf{I}. \tag{2.30}$$

The received signal $\mathbf{r}_k$ is used to derive an estimate $\hat{s}_k$ of the transmitted symbol s_k. The performance of this estimation can be computed by a cost function $\phi_k(\mathbf{f}_k, \mathbf{W})$. Examples of such cost functions are

$$\phi_k(\mathbf{f}_k, \mathbf{W}) = \begin{cases} \text{MSE}_k &= |\mathbf{f}_k^H\mathbf{Hw}_k - 1|^2 + \mathbf{f}_k^H\mathbf{R}_k\mathbf{f}_k \\ \text{SINR}_k &= \dfrac{|\mathbf{f}_k^H\mathbf{Hw}_k|^2}{\mathbf{f}_k^H\mathbf{R}_k\mathbf{f}_k}. \end{cases} \tag{2.31}$$

This enables us to state the general MIMO transceiver problem as an optimization of the sum cost function by appropriately choosing the precoder and received filters [14], that is,

$$\max_{\mathbf{W},\mathbf{F}} \sum_k \phi_k(\mathbf{f}_k, \mathbf{W})$$

$$s.t.\,\text{Tr}\,(\mathbf{WW}^H) = P_{\text{Total}}. \tag{2.32}$$

We now consider some special cases in the following examples.

Example 2.5.4 *Consider a SIMO channel i.e. $n_T = 1$, given by*

$$\mathbf{y} = \mathbf{h}s + \mathbf{z}, \tag{2.33}$$

where s is the transmitted data symbol. Let the received filter be $\mathbf{f}$. The expression for SNR is

$$\text{SNR} = \frac{|\mathbf{f}^H\mathbf{h}|^2}{\mathbf{f}^H\mathbf{f}}. \tag{2.34}$$

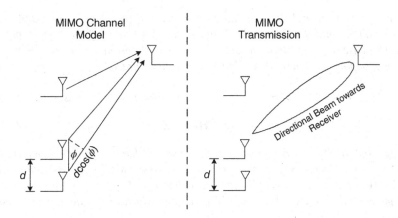

Figure 2.15 MIMO beamforming for a MISO LoS Channel

From Cauchy-Schwarz's inequality, $|\mathbf{f}^H\mathbf{h}| \leq |\mathbf{f}||\mathbf{h}|$ and hence the SNR is maximized when $|\mathbf{f}| \propto |\mathbf{h}|$. This means that the receive filter is matched to the channel and is called the matched filter. *The receiver is said to achieve* maximum ratio combining *or* receive beamforming gain.

Example 2.5.5 *Consider a MISO channel i.e. $n_R = 1$, given by*

$$y = \mathbf{h}^H\mathbf{w}s + z \tag{2.35}$$

with the power constraint $\mathbf{w}^H\mathbf{w} \leq P_{\text{Total}}$. The SNR is $|\mathbf{h}^H\mathbf{w}|^2/\sigma^2$. It can be shown that the SNR is maximized by choosing $\mathbf{w} = (\mathbf{h}/|\mathbf{h}|)\sqrt{P_{\text{Total}}}$. This is called transmit beamforming as the transmission is along the eigendirections of the spatial channel matrix and the word beam *is used to signify this directional transmission. Note that in LTE, in addition to the total power constraint each individual transmit signal from each antenna needs to be of the same strength (constant modulus property). Hence $|w_1| = |w_2| = \cdots = |w_{n_T}|$. With this additional constraint it can be shown that the optimal precoder is $w_k = \sqrt{P_{\text{Total}}/n_T}\,\theta_k$, where $\theta_k = h_k/|h_k|$ is the phase of the individual channel link.*

Example 2.5.6 *Consider a special case of Example 2.5.5 where the MISO channel is a line of sight (LoS) channel with no fading. This is shown in Figure 2.15. The SIMO channel is given by*

$$\mathbf{h} = a\mathrm{e}^{\left(j\frac{2\pi D}{\lambda}\right)}[1, \mathrm{e}^{-j2\pi d\cos(\phi)/\lambda}, \cdots, \mathrm{e}^{-j2(n_T-1)\pi d\cos(\phi)/\lambda}]^T, \tag{2.36}$$

where a is the attenuation, d is the spacing between the antenna elements, D is the separation of the transmitter and receiver and λ is the wavelength. We define $\theta = j2\pi d\cos(\phi)/\lambda$. The transmit precoder, normalized to unity power, is then

$$\mathbf{w} = \frac{1}{n_T}[1, \mathrm{e}^{\theta}, \cdots, \mathrm{e}^{(n_T-1)\theta}]^T. \tag{2.37}$$

Such a design is also called Grid of Beams *in LTE terminology. From the figure we can see that the spatial energy of the resultant transmitted radio signal is confined in a narrow angular region around the LoS angle between the transmitter and receiver, i.e. the transmit signal is a narrow beam around the LoS angle. In this case the meaning of the term beamforming is obvious. This case is of importance in LTE as sometimes the estimates of angle of arrival (AoA) and angle of departure (AoD) are used to design optimal precoders.*

Example 2.5.7 *Consider a MIMO channel where a single layer is transmitted, that is,*

$$\mathbf{y} = \mathbf{H}\mathbf{w}s + \mathbf{z}. \tag{2.38}$$

Following a similar SNR-based approach it can be shown that the optimal beamforming vector is $\mathbf{w} = \sqrt{P_{\text{Total}}}\mathbf{v}_{\max}$, where $\mathbf{v}_{\max}$ is the eigenvector corresponding to the maximum eigenvalue $\lambda_{\max}$ of $\mathbf{H}^H\mathbf{H}$. It can be shown that the receive SNR is $\lambda_{\max}^2 P_{\text{Total}}/\sigma^2$. This is of special interest in LTE and is called the closed loop rank 1 precoding *which will be dealt with in more detail in Chapter 3.*

In all the examples shown only a single layer was being transmitted. For this case, a beamforming-based transmit precoder or receive filter maximized the SNR. If multiple layers are transmitted then they interfere at the receiver. A suitable design criterion would be to minimize the mean square error as shown in Equation (2.31). A minimum mean square error estimator (MMSE) receiver minimizes the mean square error and is given by

$$\mathbf{F}^H = (\mathbf{H}\mathbf{W})^H[(\mathbf{H}\mathbf{W})(\mathbf{H}\mathbf{W})^H + \sigma^2\mathbf{I}]^{-1} \tag{2.39}$$

$$= [(\mathbf{H}\mathbf{W})^H(\mathbf{H}\mathbf{W}) + \sigma^2\mathbf{I}]^{-1}(\mathbf{H}\mathbf{W})^H. \tag{2.40}$$

If inter-cell interference is also to be considered while taking into account the design of the MMSE filter, the covariance of the out-of-cell interference $\mathbf{R}_{\text{out}}$ should be added to the $\sigma^2\mathbf{I}$ term. This quantity is unknown to the UE. The eNodeB serving this UE can cooperate with the interfering eNodeBs and inform the UE about $\mathbf{R}_{\text{out}}$: either an estimate of the instantaneous value or the long-term average power.

For multiple layers, the base station has to allocate its transmit power among the various layers. It can follow the waterfilling approach which was mentioned in Section 2.5.2 or, for simplification, allocate the power uniformly to all layers.

Note that the MMSE is the best linear receiver for minimizing MSE. The performance of MMSE can be further improved by non-linear techniques such as successive interference cancellation (SIC) at the receiver. In this case, the receiver decodes the symbol in the first layer in the presence of interference from all other layers. The received signal component due to the first symbol is estimated and then subtracted from the received signal. The second layer is then decoded; the process is repeated for all subsequent layers. This is called a MMSE-SIC receiver.

2.5.2.2 Diversity

Recall that in spatial multiplexing the transmitter is aware of the channel matrix and hence can perform beamforming along its eigendirections. The transmitter in LTE usually relies on feedback from the receiver to obtain the estimate of the channel matrix. This feedback may not be reliable due to errors in the feedback channel. Alternatively, the UE may have high mobility due to which the channel changes rapidly making it difficult to estimate. In such cases the operating principle of the transmitter should be to increase the reliability of the transmitted layers. We illustrate this with two examples.

Example 2.5.8 *Consider a 2×1 MISO system with channel $\mathbf{h} = [h_1, h_2]^T$ and a data symbol s to be transmitted. When $\mathbf{h}$ is known at the transmitter, the achievable SNR at the receiver is $\mathrm{SNR}_0 = (|h_1|^2 + |h_2|^2)P/\sigma^2$ via transmit beamforming. This has been illustrated in Example 2.5.5. Now consider that the channel is not known at the transmitter. Beamforming is not possible and if the transmitter transmits s over both antennae, then the signal vectors (rather than their powers) add up at the receiver.*

We now show that the transmitter can still ensure that SNR_0 is achieved at the receiver. For this the transmitter extends its transmission over two time slots. We make the assumption that the channel does not vary across the slots. In the first time instant, s is transmitted from the first antenna leading to a received $\mathrm{SNR}_1 = |h_1|^2 P/\sigma^2$, and in the second time instant s is transmitted from the second transmitter leading to $\mathrm{SNR}_2 = |h_2|^2 P/\sigma^2$. SNR_0 can be obtained by combining the signals across the two time slots.

The above example achieved the same SNR as transmit beamforming with channel knowledge at transmitter. Since one symbol s was transmitted over two time slots however, the transmit rate is halved as compared to transmit beamforming. Can the rate and SNR be kept the same even without transmitter knowledge? This indeed is possible by a scheme known as *Alamouti coding* [15].

Consider two time slots over which we will attempt to transmit two data symbols s_1 and s_2. Let $x_p(k)$ denote the transmitted symbol from antenna p at time k. At the first time instant, define $x_1(1) = s_1$ and $x_2(1) = s_2$ and at the second time instant define $x_1(2) = -s_2^*$ and $x_2(2) = s_1^*$. The received signal signals over the two time slots are

$$[y(1), y(2)] = [h_1, h_2] \begin{vmatrix} s_1 & -s_2^* \\ s_2 & s_1^* \end{vmatrix} + [z_1, z_2] \tag{2.41}$$

which can be rewritten

$$\begin{bmatrix} y(1) \\ y(2)^* \end{bmatrix} = \begin{bmatrix} h_1 & h_2 \\ h_2^* & -h_1^* \end{bmatrix} \begin{bmatrix} s_1 \\ s_2 \end{bmatrix} + \begin{bmatrix} z_1 \\ z_2^* \end{bmatrix}. \tag{2.42}$$

Observe that this is an equivalent MIMO transmission model with the transmit symbol vector $\mathbf{s} = [s_1, s_2]^T$. The channel matrix in Equation (2.42) has orthogonal columns and hence the detection problem for s_1 and s_2 can be decomposed into two parallel channels each having $\mathrm{SNR} = \mathrm{SNR}_0 = (|h_1|^2 + |h_2|^2)P/\sigma^2$.

Alamouti coding therefore achieves the same SNR (diversity order) and rate without channel knowledge at the transmitter. However, the Alamouti scheme is designed for the $N_T = 2$ case only and does not extend to more than two transmit antennae. For more than two transmit antennae, more general space-time codes (or space-frequency codes when transmission is over two orthogonal subcarriers instead of two time slots) have been designed to achieve full diversity order. However, they incur some losses in achievable rate.

Due to the presence of multiple orthogonal subcarriers in OFDM structure, Space Frequency Block Codes (SFBC) have been adopted in LTE. Transmission schemes similar to Alamouti have been designed by exploiting the diversity over different subcarriers.

Diversity schemes such as SFBC have reduced UE feedback and downlink control signaling overhead requirements compared to spatial multiplexing and are sometimes favored in low-rate applications such as voice over IP (VoIP) [16].

2.5.2.3 Multi-User MIMO

In the LTE downlink, MU-MIMO is simultaneous transmission to multiple UEs from the same eNodeB. This is an important method to increase system capacity. Channel models seen in realistic LTE deployments are often highly correlated and hence each UE has a small channel rank, r. The rank r of a UE determines the number of independent layers that may be transmitted to it. If the eNodeB has more antennae than the number of layers that can be transmitted to one UE, it can transmit to another UE simultaneously over the same set of resources. The layers intended for one UE interfere with the transmissions of the other co-scheduled UEs. The nature of the interference is similar to the inter-layer interference when multiple layers are transmitted to the same UE. However, since the UE receivers cannot cooperate and design a common MMSE receiver for all layers, the eNodeB has to ensure that the resulting inter-UE interference is minimized.

Let us consider the MU-MIMO scenario with two UEs indexed 1 and 2. Let the channels and transmit precoders for the UEs be $\mathbf{H}_1, \mathbf{W}_1$ and $\mathbf{H}_2, \mathbf{W}_2$ respectively and the received signals be $\mathbf{y}_1$ and $\mathbf{y}_2$. These are given by

$$\mathbf{y}_1 = \mathbf{H}_1(\mathbf{W}_1\mathbf{s}_1 + \mathbf{W}_2\mathbf{s}_2) + \mathbf{z}_1 \tag{2.43}$$

$$\mathbf{y}_2 = \mathbf{H}_2(\mathbf{W}_1\mathbf{s}_1 + \mathbf{W}_2\mathbf{s}_2) + \mathbf{z}_2. \tag{2.44}$$

The interfering signal at $\mathbf{y}_1$ is $\mathbf{H}_1\mathbf{W}_2\mathbf{s}_2$ and that at $\mathbf{y}_2$ is $\mathbf{H}_2\mathbf{W}_1\mathbf{s}_1$. The precoders can be designed such that $\mathbf{H}_1\mathbf{W}_2 = 0$ and $\mathbf{H}_2\mathbf{W}_1 = 0$. This is called *block diagonalization* [17]. This precoder design is optimized to reduce interference to the co-scheduled UE. Block diagonalization works best if the two channels $\mathbf{H}_1$ and $\mathbf{H}_2$ are orthogonal or as close to orthogonal as possible (i.e. the eigendirections of the two channels are perpendicular to each other). In this case, nulling out the interference of the co-scheduled UE also amounts to transmission along the eigendirection of the intended channel of the UE. In other words, $\mathbf{W}_1$ is aligned to the eigendirection of the channel correlation matrix $\mathbf{H}_1^H\mathbf{H}_1$ and moreover $\mathbf{H}_2\mathbf{W}_1 \sim 0$. A LTE eNodeB typically has multiple associated UEs and it is the job of the scheduler at MAC layer to pick a UE pair for MU-MIMO transmission so that block diagonalization performs optimally.

References

[1] 3GPP (2012) Evolved Universal Terrestrial Radio access (E-UTRA) and Evolved Universal Terrestrial Radio Access (E-UTRAN), overall description 3GPP TS 36.300 v11.2.0. Third Generation Partnership Project, Technical Report.

[2] Sesia, S., Toufik, I., and Baker, M. (2011) *LTE: The UMTS Long Term Evolution: From Theory to Practice*. United Kingdom: John Wiley & Sons Ltd.

[3] Dahlman, E., Parkvall, S., and Skold, J. (2011) *4G: LTE/LTE-Advanced for Mobile Broadband*. United Kingdom: Academic Press.

[4] UMTS (2010) UTRAN overall description, ETSI TS 125 401 v9.0.0. European Telecommunications Standards Institute, Technical Report.

[5] 3GPP (2013) 3GPP TR 36.932, scenarios and requirements for small cell enhancements for E-UTRA and E-UTRAN. Third Generation Partnership Project, Technical Report.

[6] Orfanidis, S.J. (1995) *Introduction to Signal Processing*. New Jersey: Prentice-Hall, Signal Processing Series.

[7] Rappaport, T. (1996) *Wireless Communications: Principles and Practice*. New Jersey: Prentice Hall.

[8] Huang, J., Subramanian, V., Agrawal, R., and Berry, R. (2006) Downlink scheduling and resource allocation for OFDM systems. In *Proceedings of IEEE Conference on Information Sciences and Systems (CISS)*, March 2006.

[9] Foschini, G. and Gans, M. (1998) On limits of wireless communications in a fading environment when using multiple antennas. *Wireless Personal Communications*, **6**, 311–335.

[10] Li, Q., Li, G., Lee, W., Lee, M., Mazzarese, D., Clerckx, B., and Li, Z. (2010) MIMO techniques in WiMAX and LTE: a feature overview. *IEEE Communications Magazine*, **48**(5), 86–92.

[11] Zheng, L. and Tse, D. (2013) Diversity and multiplexing: A fundamental tradeoff in multiple antenna channels. *IEEE Transactions on Information Theory*, **49**, 1073–1096.

[12] Telatar, I. (1999) Capacity of multi-antenna gaussian channels. *European Transactions on Telecommunications*, **10**, 585–595.

[13] Tse, D. and Viswanath, P. (2005) *Fundamentals of Wireless Communication*. United Kingdom: Cambridge University Press.

[14] Palomar, D. and Jiang, Y. (2006) *MIMO Transceiver Design via Majorization Theory*. Now Publishers, Foundations and Trends in Communications and Information Theory, 3(4–5).

[15] Alamouti, S. (1998) A simple transmit diversity technique for wireless communications. *IEEE Journal on Selected Areas in Communications*, **16**, 1451–1458.

[16] Dahlman, E., Furuskar, A., Jading, Y., Lundevall, M., and Parkvall, S. (2008) Key features of the LTE radio interface. *Ericsson Review*, **2**), 77–80.

[17] Shen, Z., Chen, R., Andrews, J., Heath, R. and Evans, B. (2007) Sum capacity of multiuser MIMO broadcast channels with block diagonalization. *IEEE Transactions on Wireless Communications*, **6**(6), 886–890.

3

LTE Signal Structure and Physical Channels

3.1 Introduction

Orthogonal Frequency Division Multiple Access (OFDMA) and Single-Carrier FDMA (SC-FDMA) are employed in the downlink and uplink of LTE, respectively, both of which use OFDM. In this chapter, we introduce the OFDM-based LTE signal and frame structure to explain how the user-plane and control-plane data from the higher layers are multiplexed with physical layer control and reference signals (RSs) to generate physical layer signals for transmission. We discuss the LTE signal structure in Section 3.2. In Section 3.3, we briefly review LTE downlink and uplink operations by introducing physical layer channels as well as various RSs and show that they are mapped to higher-layer channels. The detailed functionality of the physical channels and how they are mapped to the time frequency grid of OFDM symbols are explained in Section 3.5 and Section 3.6 for downlink and uplink, respectively.

3.2 LTE Signal Structure

The LTE signal structure spans the three dimensions of space, frequency, and time [1]. In the spatial dimension, multiple independent data streams can be effectively transmitted by using MIMO at the eNodeB and UE. In the frequency dimension, transmission over multiple subcarriers is made possible by OFDM. In the time dimension, consecutive transmitted OFDM symbols are grouped into slots of duration 0.5 ms. The largest unit of time domain transmission is the 10 ms radio frame, which is further divided into ten 1 ms subframes. The duration of 1 ms is also known as Transmission Time Interval (TTI). Each subframe consists of two slots.

LTE has two duplexing modes: Frequency Division Duplexing (FDD) and Time Division Duplexing (TDD). In FDD mode, a paired spectrum of two separate and equal-sized bands are used, one each for the uplink and the downlink transmission. All subframes in each radio frame for FDD have the same structure. In TDD mode, the subframes of each radio frame are divided into two subsets, one each for downlink and uplink transmissions. The choice

Heterogeneous Networks in LTE-Advanced, First Edition. Joydeep Acharya, Long Gao and Sudhanshu Gaur.
© 2014 John Wiley & Sons, Ltd. Published 2014 by John Wiley & Sons, Ltd.
Companion Wesite: www.ltehetnet.com.

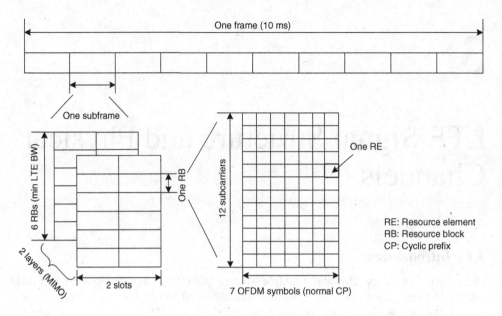

Figure 3.1 LTE frame structure in FDD mode

of duplexing mode by an operator depends on spectrum availability in a region and also government regulations.

The detailed structure of a radio frame in FDD mode is shown in Figure 3.1. Each subframe can support multiple spatial MIMO layers. Each 0.5 ms slot in a subframe has seven OFDM symbols if OFDM with normal control plane (CP) length (5 μs) is used and six OFDM symbols if extended CP (17 μs) is used. For purposes of resource allocation, the LTE signal is divided in time and frequency into units called Physical Resource Block (PRB). In frequency, a PRB spans 12 subcarriers each with spacing 15 kHz; in time, it is of the duration of a slot. A PRB is further divided into Resource Elements (REs). One RE is one OFDM subcarrier for the duration of one OFDM symbol and is the smallest indivisible unit in the LTE time frequency grid. A PRB therefore comprises 84 REs in the case of the normal CP and 72 REs in the case of the extended CP. Different sets of REs, which are mutually exclusive of each other, are reserved for different physical channels and RSs. These physical channels and RSs will be explained in detail in the next few sections.

LTE supports a variety of bandwidths up to 20 MHz. The number of PRBs for different LTE bandwidths is listed in Table 3.1. The minimum LTE bandwidth is 1.4 MHz (6 PRBs) and this is shown in Figure 3.1.

The detailed structure of a radio frame in TDD mode is shown in Figure 3.2. The subframes with 'D' and 'U' are reserved for downlink and uplink transmissions, respectively. These subframes have the same structure as the corresponding downlink and uplink subframes in FDD mode. The subframes with 'S' are special subframes with three fields: the Downlink Pilot Time Slot (DwPTS), Guard Period (GP) and Uplink Pilot Time Slot (UpPTS). The DwPTS can be treated as a normal downlink subframe with reduced length, and thus the amount of data which can be transmitted in the DwPTS is less. The field GP is used to protect switching between downlink to uplink transmissions. The UpPTS has a very short duration and is not used for

Table 3.1 Number of PRBs for different LTE bandwidths

Channel bandwidth (MHz)	Number of PRBs
1.4	6
3	15
5	25
10	50
15	75
20	100

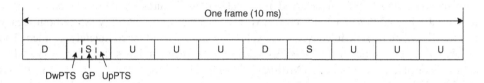

Figure 3.2 LTE frame structure in TDD mode

uplink transmission. It can be used for channel sounding, random access or an extra Guard Period between downlink and uplink transmissions.

The frame structure in TDD has seven configurations as listed in Table 3.2. Each configuration has a different arrangement of the downlink, uplink, and special subframes in a radio frame. Some configurations have more subframes for downlink transmission while others have more for uplink. Depending on downlink and uplink traffic loads, the network operator can choose the configuration with the optimal downlink–uplink subframe ratio. This flexibility is not possible in FDD; however, TDD requires tighter synchronization. Note that the special subframe occurs whenever there is a switch from a downlink to an uplink subframe.

In Release 9, a new type of subframe has been introduced for Multimedia Broadcast/ Multicast Service (MBMS) operations. This is known as the Multimedia Broadcast Single Frequency Network (MBSFN) subframe. In a MBSFN subframe, multiple eNodeBs transmit the

Table 3.2 TDD downlink–uplink configuration

Config. #	Subframe index									
	#0	#1	#2	#3	#4	#5	#6	#7	#8	#9
0	D	S	U	U	U	D	S	U	U	U
1	D	S	U	U	D	D	S	U	U	D
2	D	S	U	D	D	D	S	U	D	D
3	D	S	U	U	U	D	D	D	D	D
4	D	S	U	U	D	D	D	D	D	D
5	D	S	U	D	D	D	D	D	D	D
6	D	S	U	U	U	D	S	U	U	D

(Reproduced by permission of © 3GPP.)

same signal simultaneously so that the received signal to noise ratio increases. A UE scheduled for reception in a MBSFN subframe will receive multiple versions of the signal coming from multiple eNodeBs with different propagation delays. This appears as additional multipaths to the UE and thus the overall delay spread increases. Since the cyclic prefix length has to be greater than the overall delay spread, only the extended CP is used for MBSFN transmission.

Another important use of MBSFN subframes has to do with the issue of backward compatibility. To understand this let us recollect how LTE standardization works. Initially LTE Release 8 introduced a certain set of transmission features and associated signaling that the Release 8 UEs and eNodeBs would be capable of understanding. In subsequent releases, new features such as eICIC (LTE Release 10) and CoMP (LTE Release 11) have been introduced which will benefit new UEs that have been designed to exploit these features. However, the corresponding signaling from the eNodeB to enable these features is not *understood* by legacy Release 8 UEs and this can cause problems with backward compatibility. Since both legacy and new UEs can be scheduled in normal subframes, this cannot be completely avoided. The legacy UEs are not scheduled in MBSFN subframes however, and they do not try to decode a subframe once it detects that it is a MBSFN subframe. Thus if the eNodeB schedules new UEs in a MBSFN subframe and uses the newer advanced transmission schemes such as Enhanced Inter-cell Interference Coordination (eICIC) or Coordinated Multi-point Transmission Reception (CoMP), no interference is created at the legacy UE [2].

Certain subframes in a radio frame are not allowed to be configured as MBSFN subframes. These are the subframes that carry Synchronization Signals (SSs) and broadcast system information. Particularly, subframes 0^1, 4, 5, and 9 in each radio frame are not allowed in FDD mode, while subframes 0, 1, 2, 5, and 6 cannot be configured as MBSFN subframes in TDD mode.

3.3   Introduction to LTE Physical Channels and Reference Signals

In this section, we introduce the LTE channels and physical layer RSs. In particular, we explain how the control-plane and user-plane data from the higher layers is mapped on the physical channels and multiplexed with the RSs for data transmission. We start with the LTE downlink and then discuss the uplink case. We present only a brief overview in this chapter; for more details refer to [1,3–5].

In the LTE downlink, the physical channels can be categorized into two groups. The first is called *Data Channels* as they are used to transport information that is generated at layers above the physical layer. Different examples of data channels are as follows.

- **Physical Broadcast Channel (PBCH):** carries the master information block (MIB) which has important system information needed by the UEs for initial access and configuration to the network (see Section 3.5.1).
- **Physical Downlink Shared Channel (PDSCH):** carries user data and the system information that are not carried by the PBCH (see Section 3.5.2).
- **Physical Multicast Channel (PMCH):** carries data for Multimedia Broadcast and Multicast Services (see Section 3.5.3).

---

[1] The subframes of a radio frame are indexed from 0 to 9.

The second category of physical channels are called *Control Channels* as the information they carry is generated at the physical layer and is used mostly for carrying control information to support the data channel transmission.

- **Physical Control Format Indicator Channel (PCFICH):** informs the UE of the number of OFDM symbols used for PDCCH which is the control region (see Section 3.5.4).
- **Physical Hybrid ARQ Indicator Channel (PHICH):** carries Hybrid ARQ ACK/NAKs in response to the uplink transmissions (see Section 3.5.5).
- **Physical Downlink Control Channel (PDCCH):** informs the UE of the resource allocation and carries the uplink scheduling grant (see Section 3.5.6).

Downlink reference signals are used by the UE for cell synchronization and channel estimation purposes. LTE has the following downlink RSs, discussed in more detail in Chapter 4.

- **Synchronization Signal (SS):** used for initial downlink synchronization to the network.
- **Cell-Specific RS (CRS):** used for downlink channel estimation and/or demodulation of downlink data signals.
- **Channel State Information (CSI) RS:** introduced in Release 10 for downlink channel estimation.
- **UE-Specific RS associated with PDSCH:** introduced in Release 9 for demodulation of downlink data signals.

The information that is transmitted in the physical layer data channels originates in the RLC layer. The RLC layer has different buffers for different types of application data (for e.g. video/voice/NAS control messages, etc.) and these are passed to the MAC by logical channels. The MAC decides the characteristics of the data to be transported (such as channel coding parameters) and exchanges this information with the physical layer through transport channels.

To understand the LTE physical downlink channels, let us first introduce the downlink logical channels. Some of these, listed below, are used to carry the control plane data.

- **Broadcast Control Channel (BCCH):** carries the broadcasting system control information.
- **Paging Control Channel (PCCH):** carries the paging information and system information change notifications.
- **Common Control Channel (CCCH):** carries the common control information for a UE without a Radio Resource Control (RRC) connection.
- **Multicast Control Channel (MCCH):** carries the MBMS control information.
- **Dedicated Control Channel (DCCH):** carries the dedicated control information for a UE with RRC connection.

Other logical channels are used for carrying the user-plane data, as follows.

- **Dedicated Traffic Channel (DTCH):** carries UE data.
- **Multicast Traffic Channel (MTCH):** carries data for MBMS.

The CCCH, DCCH, and DTCH are bi-directional logical channels used for both downlink and uplink. The relationship between the physical channels and logical channels is shown in

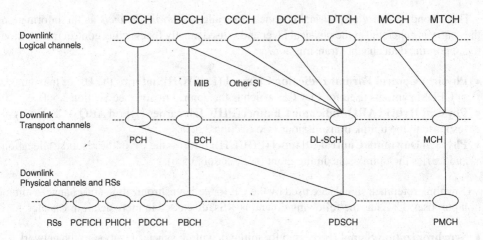

Figure 3.3 The mapping between downlink physical channels and logical channels. (Reproduced by permission of © 3GPP.)

Figure 3.3, where the logical channels are mapped to the physical channels via the transport channels. From Figure 3.3, we can make the following observations.

- The PDSCH and PMCH transport channels are almost all control-plane and user-plane data (except for the MIB which is transported by the PBCH).
- The PCFICH, PHICH, and PUCCH are not mapped to any logical or transport channels, which means they do not transport any data from the higher layers. As shown later in Section 3.5, these channels are generated in the physical layer in order to provide necessary support for the transmissions of the PBCH, PDSCH, and PMCH.
- The downlink RSs are not mapped to any logical or transport channels; they are therefore also generated at the physical layer.

For the uplink, the physical channels are also categorized into two groups depending on whether or not they carry the user-plane and data-plane data from the higher layers. The single data channel is:

- **Physical uplink shared channel (PUSCH):** carries user data (see Section 3.6.1).

The control channels are:

- **Physical uplink control channel (PUCCH):** carries the scheduling request from the UE, the HARQ acknowledgement for the downlink transmission, and the channel state information (CSI) feedback (see Section 3.6.2).
- **Physical random access channel (PRACH):** is used for initial network access and uplink time synchronization (see Section 3.6.3).

The following RSs are employed to support the uplink physical channels. These RSs are used by the eNodeB for synchronization and channel estimation; further details are provided in Chapter 4.

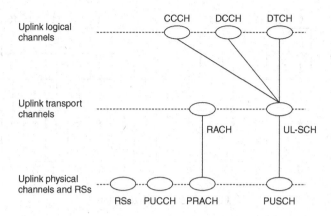

Figure 3.4 The mapping between uplink physical channels and logical channels. (Reproduced by permission of © 3GPP.)

- **Demodulation RS:** supports the transmissions of the PUCCH and PUSCH.
- **Sounding RS:** supports the transmissions of the PUSCH.

The relationship between the uplink physical channels and logical channels is shown in Figure 3.4. The logical channels are mapped to the physical channels via the transport channels, similar to the downlink case. We can see that only the PUSCH is used to transport the control-plane and user-plane data. The PUCCH and PRACH are not mapped to any logical channels, which means that they do not transport any data from the RLC layer.

3.4   Resource Block Assignment

The most straightforward way for the eNodeB to inform each scheduled UE about the RB assignment is to send a bitmap whose size is equal to the number of RBs in the system. The bit corresponding to a RB is set to 1 if that RB is allocated to the UE. This provides the maximum flexibility but the length of the bitmap increases with the system bandwidth. For example, a system bandwidth of 20 MHz (i.e. 100 RBs) will require 100 bits for the bitmap, which will incur a significant overhead especially for small packets. Different approaches for indicating the RB assignment to a UE are therefore specified in LTE. These are also called resource allocation types and require a smaller number of bits while keeping sufficient allocation flexibility. Before these approaches are explained in further detail, we first introduce Virtual Resource Blocks (VRBs).

Virtual resource blocks are used to enable distributed transmission in the frequency domain for the downlink. If a UE is allocated a pair of consecutive VRBs, these blocks need not map to consecutive PRBs. VRB to PRB mapping is fixed and known both at UE and base station. A VRB is of the same size as a PRB. The VRBs are used in pairs which can be categorized into two types: localized and distributed VRB pairs. A localized VRB pair is directly mapped to a localized PRB pair as shown in Figure 3.5. In this case, the VRB pair with index n is mapped to the localized PRB pair with the same index. In the case of the distributed VRBs, a VRB pair is mapped to a distributed PRB pair as shown in Figure 3.6 based on the following two criteria:

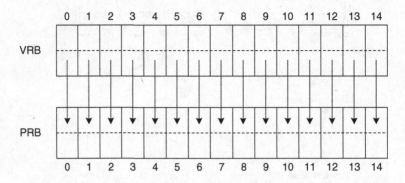

Figure 3.5   Localized VRB

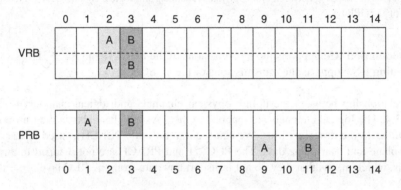

Figure 3.6   Distributed VRB

- consecutive VRB pairs are not mapped to consecutive PRB pairs in frequency domain; and
- the two RBs of each distributed PRB pair are transmitted with a certain gap in frequency domain.

In LTE, the localized VRBs are used in resource allocation types 0 and 1, while the distributed VRBs are used in resource allocation type 2 to facilitate distributed PRB assignment. We explain these three resource allocation types in detail in the following.

Resource Allocation Type 0

In resource allocation type 0, the RBs are grouped (RBGs) according to their locations in the frequency domain, and a bitmap is used to indicate the assigned RBGs to a UE. The RBGs have the same size P, which is a function of the system bandwidth as shown in Table 3.3. Let N_{RB}^{DL} denote the downlink system bandwidth in terms of RBs. The length of the bitmap for resource allocation type 0 is given by $N_{RBG} = \lceil N_{RB}^{DL}/P \rceil$, where $\lceil \cdot \rceil$ denotes the ceiling operation. An example for the case of $N_{RB}^{DL} = 15$, $P = 2$, and $N_{RBG} = 8$ is given in Figure 3.7, where each bit in the bitmap indicates a pair of consecutive RBs in the frequency domain. Note that only the localized RB pairing (as explained in Section 3.5.2) is used in resource allocation 0.

Table 3.3 RBG size for resource allocation type 0/1

System bandwidth	RBG size
0–10	1
11–26	2
27–63	3
64–110	4

(Reproduced by permission of © 3GPP.)

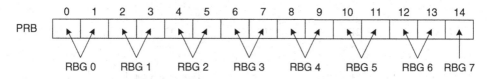

Figure 3.7 Resource allocation type 0

Resource Allocation Type 1

In resource allocation type 1, the RBGs are formed as in resource allocation type 0. The RBGs are further divided into P RBG subsets. A RBG subset p, where $0 \leq p < P$, consists of every Pth RBG starting from RBG p. A bitmap is used to indicate the RBGs assigned to a UE in a particular RBG subset. Since the eNodeB also needs to send the RBG subset index, the bitmap in resource allocation type 1 is smaller than that in resource allocation type 0 in order to maintain the same total number of bits used for RB assignment. As a result, the bitmap does not have enough bits to indicate all RBGs in a RBG subset. To resolve this issue, an extra bit is used to indicate whether the bitmap relates to the 'left' or 'right' part of the RBGs in a RBG subset. An example for the case of $N_{RB}^{DL} = 15$, $P = 2$, and $N_{RBG} = 8$ is given in Figure 3.8, where the RBGs are divided into two subsets and a bitmap of length 6 is used to indicate the assigned RBs in a particular subset.

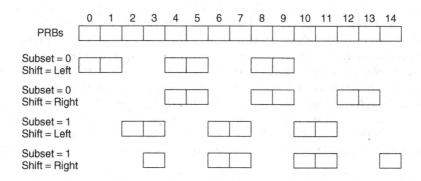

Figure 3.8 Resource allocation type 1

Table 3.4 Resource allocation types for each DCI format

DCI format	Resource allocation types
0	Type 0 or Type 1
1	Type 0 or Type 1
1A	Type 2
1B	Type 2
1C	Type 2
1D	Type 2
2	Type 0 or Type 1
2A	Type 0 or Type 1
2B	Type 0 or Type 1
2C	Type 0 or Type 1
2D	Type 0 or Type 1
3	None
3A	None
4	Type 0 or Type 1

Resource Allocation Type 2

In resource allocation type 2, the RB assignment information indicates a contiguous set of VRBs which could be localized or distributed. The RB assignment can vary from a single RB up to the maximum available RBs. The starting RB index and the length of the assigned RBs are signalled to the UE.

Since different resource allocation types have different parameters, the format of the control signal used to support them should also be different. The format and content of a control signal is known as DCI (Downlink Control Information). For example, DCI Format 1A can only support resource allocation type 2. The resource allocation types supported by each DCI format are summarized in Table 3.4

3.5 Downlink Physical Channels

As explained in Section 3.3, the downlink physical channels can be categorized into two groups: data channels (PBCH, PDSCH, and PMCH) that carry the user-plane and control-plane data from the higher layers, and control channels (PCFICH, PHICH, and PDCCH) that support the data channels. In Figure 3.9, a high-level picture is provided to show how these channels are mapped to the LTE resource grid. In each subframe, the first N OFDM symbols are used to carry the physical control channels. In the subframes for non-MBMS service[2], the value of N can be chosen from $\{1, 2, 3, 4\}$. This size of the control channel is therefore variable and depends on the amount of control information to be transmitted. This in turn depends on the number of UEs that are scheduled for transmission in the subframe and the transmission mode for each UE (e.g. if a UE has been scheduled with MIMO it needs more associated control information than a UE scheduled with SISO). The design of LTE thus optimizes the REs

---

[2] These subframes could be either non-MBSFN subframes or Release 10/11 MBSFN subframes, which are configured for unicast transmissions.

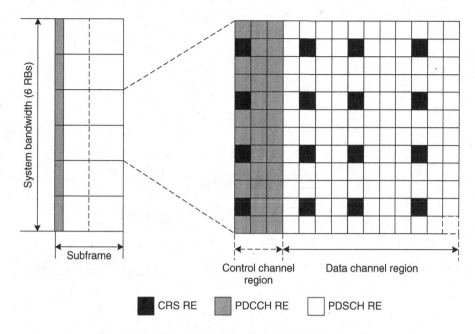

Figure 3.9   Control channel region and data channel region

available for data transmission via the variable length control channel design. In the subframes for MBMS transmission, a smaller N is chosen from $\{1, 2\}$. This is because the resource allocation information is semi-static and common for all UEs in MBMS subframes, which is signalled by higher-layer signaling rather than PDCCH signaling. The physical data channels are located in the data channel region, which occupies the remaining part of the subframe.

The CRS and other RSs listed in Section 3.3 are embedded in the control channel and data channel regions. In the following sections, we explain each of the downlink physical channels in detail.

3.5.1   Physical Broadcast Channel (PBCH)

The PBCH carries the MIB which is the essential system information for initial access to a cell. The MIB consists of the downlink system bandwidth (3 bits), PHICH size (3 bits), the system frame number (SFN) (8 bits), and 10 spare bits. The MIB is repeated every 40 ms, which translates to a data rate of just 350 bps on the PBCH. Since reading the MIB is vital for initial cell detection and acquisition, reliability is the most important issue while designing the PBCH structure. In order to guarantee reliable reception of the PBCH, the following three mechanisms are utilized: forward error correction (FEC) coding, antenna diversity, and time diversity.

The detailed PBCH processing is shown in Figure 3.10. After adding 16 bits CRC, the convolutional coding and rate matching are applied to the MIB bit sequence. The basic coding rate for the convolutional encoder is 1/3. After the rate matching, the MIB is coded at the rate of 1/48 to have strong error protection. The coded sequence is then scrambled with a cell-specific sequence, after which Quadrature Phase Shift Keying (QPSK) is employed for

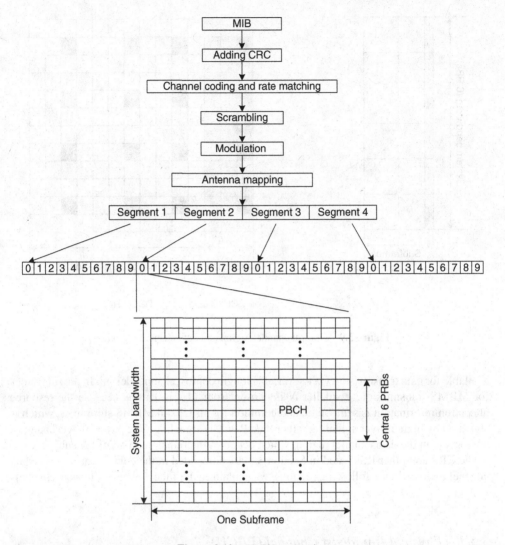

Figure 3.10 PBCH processing

the modulation. The block of the modulation symbols are then mapped to multiple antennas via Space-Frequency Block Code (SFBC) in order to exploit the antenna diversity if the eNodeB employs multiple antennae. Note that since the UE does not know the number of transmit antennae employed by the eNodeB during the initial cell acquisition phase, it performs blind decoding by assuming all possible SFBC receivers corresponding to the different possible numbers of transmit antennae.

After the antenna mapping, the block of coded symbols for each antenna is divided into four equal-sized individually self-decodable units. These units are transmitted during four consecutive radio frames to exploit the time diversity. This significantly reduces the probability of the decoding failure. If the receive signal to interference and noise ratio (SINR) is good, then the UE may be able to decode the MIB from the first unit itself. In this case the UE does not have to decode the other three units. If the SINR is low and the UE fails to decode the MIB from the first unit, the UE can wait until it receives the second unit. It then soft-combines it with the first one and tries to decode the MIB again. The UE therefore does not have to wait until all the four units have been received before it can start attempting to decode the MIB. This reduces the decoding latency and saves UE battery life.

The REs reserved for the PBCH are located at the central six RBs in the first subframe of each frame. Note that any REs which may be used for reference signals (RSs) are avoided by the PBCH. The reason why the central six RBs are used for the PBCH is that the system bandwidth information is also not available to the UE before decoding the PBCH. Regardless of the actual system bandwidth, the central six RBs (corresponding to the minimum possible bandwidth of 1.4 MHz) are always present.

3.5.2   Physical Downlink Shared Channel (PDSCH)

The PDSCH is the main data channel for the LTE downlink. It is used to transport user data, to broadcast system information (except for the MIB which is carried on the PBCH) and to send paging messages. The REs reserved for the PDSCH in each subframe[3] is shown in Figure 3.11. The starting point of the PDSCH region is configurable, which could be the 2nd, 3rd, or 4th OFDM symbol [4]. Note that any REs in the PDSCH region which are used for the PBCH and RSs will be not used for the PDSCH.

When the PDSCH is used to transport user data, the smallest resource unit allocated to a UE is one pair of PRBs, which are located in the same subframe but different time slots. There are two types of PRB pairs: localized and distributed PRB. The two PRBs for the localized pairing occupy the same set of 12 subcarriers, while the two PRBs in the distributed pairing occupy different sets of 12 subcarriers. Figure 3.11 shows an example of both localized and distributed PRB pairs. One advantage of distributed PRB pairs is frequency diversity. Consider the case of semi-static scheduling, where the same PRB pairs are allocated to a specific UE

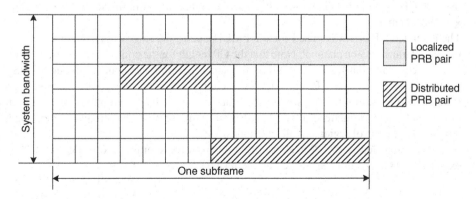

Figure 3.11   Localized and distributed PRB pairs

---

[3] From Release 10, the PDSCH can also be mapped to the data region of MBSFN subframes.
[4] The control channel region occupies at least the first OFDM symbol in LTE Release 11 and previous specifications. In Release 12, it is possible that the PDSCH starts from the first OFDM symbol.

on periodic basis. If a UE suffers from deep fading in a particular sub-band, localized PRB allocation increases the chances of both PRBs being in perpetual deep fade.

Distributed PRB pairing increases the complexity of resource allocation and is not supported in certain transmission modes that use UE-specific RS for demodulation.

The idea of transmission modes was alluded to above. In LTE the term transmission mode (TM) is used to denote the way in which data are transmitted from multiple antennae and how they are subsequently demodulated at the UE side. Different transmission modes may differ in the signal processing used at the eNodeB (spatial multiplexing versus diversity) or in the demodulation used at the UE (use of CRS- or UE-specific DMRS). The eNodeB configures different TMs for different UEs via RRC signaling. Table4.10 describes various TMs used for PDSCH transmission.

The PRB pairs in each subframe in the PDSCH region can be allocated to different UEs by the eNodeB. The allocation can be dynamic, in which case the eNodeB informs each UE of its allocated PRB pairs by dynamic PDCCH signaling. The allocation can also be semi-static in which case the resources allocated to a UE remain constant over multiple subframes. Semi-static scheduling is useful for applications such as voice that need a constant bit rate. RRC signaling is used to inform UE of the resource allocation information in case of semi-static scheduling.

The PDSCH is also used to broadcast system information blocks (SIBs) and to send paging message.

3.5.3   Physical Multicast Channel (PMCH)

The PMCH is used to carry MBMS data; the PMCH is therefore transmitted in the MBSFN subframes. Like PDSCH, the PMCH spans the entire system bandwidth. The starting OFDM symbol for PMCH is configurable, which could be the 2nd or 3rd. The main differences between the PMCH and PDSCH are as follows.

- The resource allocation information for the PMCH is signaled by RRC signaling instead of the PDCCH signaling. This makes the control channel region in the subframes containing the PMCH smaller.
- The extended CP is always used for the PMCH due to the large effective delay spread of multi-cell transmission channel. Note that the CP length for the control channel region could be different from that for PMCH within the same MBSFN subframe. If the non-MBSFN subframes use the normal CP, then the normal CP is also used for the control channel region in the MBSFN subframes. This is because the UE has to read the control channel region of each subframe no matter if it is a MBSFN subframe or not, which requires a unique CP length for the control regions in all subframes.
- The RSs embedded in the PMCH are different from those embedded in the PDSCH. For example, the CRS is not transmitted on the PMCH.

3.5.4   Physical Control Format Indicator Channel (PCFICH)

The PCFICH is used to carry the Control Format Indicator (CFI), which indicates the number of OFDM symbols used for the PDCCH transmission. As the correct decoding of CFI is very important for subsequent decoding of PDCCH, the CFI is encoded into 32 bits in

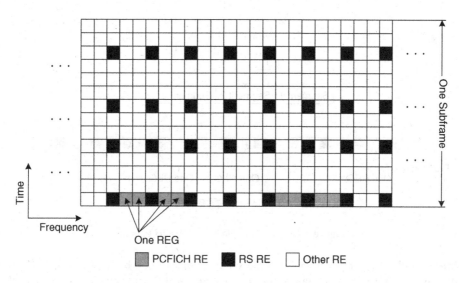

Figure 3.12 An example of a PCFICH REG

length to have sufficient error protection. These 32 bits are mapped to 16 symbols with QPSK modulation. These 16 symbols are mapped into 4 groups of 4 REs each, called a Resource Element Group (REG). These 4 REGs are transmitted in the first OFDM symbol of each subframe and distributed across the entire system bandwidth to exploit the frequency diversity. If multiple antennae are employed at the eNodeB, transmit diversity is used for the PCFICH transmission. The diversity scheme used is the same as that for transmitting the PBCH.

An example of the PCFICH REG in a downlink subframe is provided in Figure 3.12. We can see that each REG consists of 4 consecutive REs that are not used for RSs in the first OFDM symbol. Note that the location of the PCFICH is predefined for a given cell. In order to minimize the probability of collision with the PCFICH from a neighboring cell, a cell-specific frequency shift is applied to the positions of the PCFICH REGs. A UE can always locate the PCFICH based on the cell ID, which is acquired during the cell synchronization process (see Chapter 2).

3.5.5 Physical Hybrid ARQ Indicator Channel (PHICH)

The PHICH is used to carry the HARQ indicator, which informs a UE if the eNodeB has correctly received an uplink transmission. The HARQ indicator is set to 0 for an Acknowledge (ACK) and 1 for a Negative acknowledge (NACK). The 1-bit HARQ indicator is repeated three times for error protection. Each of them is then mapped to a REG of four REs.

Since multiple UEs may be scheduled simultaneously in the PUSCH RBs (see Section 3.6.1) in each subframe, multiple PHICHs may be needed (one for each scheduled UE). To avoid additional signaling to indicate which PHICH carries the HARQ indicator for a given UE, each PHICH is given an index derived from the index of the lowest uplink RB and the DMRS cyclic shift assigned to the corresponding UE. Note that if the uplink transmission occurs in subframe N, the corresponding PHICH will always be in subframe $N + 4$ and thus a UE knows where to look for its PHICH.

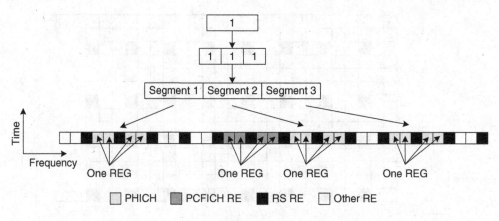

Figure 3.13   An example of the PHICH mapping

Multiple PHICHs can be mapped to the same set of REGs, which form a PHICH group. Different PHICHs within the same group are separated by different complex orthogonal Walsh sequences. The length of the Walsh sequence is 4 and 2 for the normal and extended CP cases, respectively. Since the sequence is complex, the number of PHICHs within each group can be up to twice the sequence length.

The PHICH is normally located in either the first one or three OFDM symbols in each subframe. In Figure 3.13, we show how to map the PHICH into the LTE resource grid. After the repetition coding and Walsh spreading, each segment of the PHICH is mapped to a REG in the first OFDM symbol. The PHICH mapping avoids the REs used for RSs or PCFICH.

3.5.6   Physical Downlink Control Channel (PDCCH)

The PDCCH is used to carry the Downlink Control Information (DCI), which includes downlink resource assignments, uplink grants, power control commands, and other control information for a UE or a group of UEs. The content of the DCI depends on the system and UE configurations, for example, the transmission modes on the PDSCH or PUSCH. The DCI is therefore categorized into different DCI formats. We first detail DCI Format 1A to explain the different kinds of control information that a DCI contains.

DCI Format 1A can be used in any PDSCH transmission modes, and includes the following information.

1. Carrier Indicator Field (CIF) (0 or 3 bits), which indicates the component carriers the DCI relates to. This was introduced in LTE Rel-10 Carrier Aggregation and is used if the cross-carrier scheduling is enabled via RRC signaling.
2. Flag for Format 0/1A differentiation (1 bit), where value 0 indicates DCI Format 0 and value 1 indicates DCI Format 1A. Note that the DCI Formats 0 and 1A have the same size in order to avoid additional UE complexity that would increase if the UE had to process multiple DCI formats.
3. Localized/Distributed Virtual Resource Block (VRB) assignment flag (1 bit), where value 0 indicates localized VRBs and value 1 indicates distributed VRBs. The value is set to 1 for DCI Format 1A. The concept of VRB is explained later in this section.

4. Resource block assignment, which indicates the RBs for the UE to receive the PDSCH. The size of the field depends on the cell bandwidth and the resource allocation types, discussed in detail later in this section.
5. Modulation and coding scheme (5 bits), which informs the UE about the modulation scheme, the code rate, and the transport-block size used in the PDSCH transmission.
6. HARQ process number (3 bits for FDD and 4 bits for TDD), which informs the UE of the HARQ process details.
7. New Data Indicator (1 bit), where value 1 indicates a new transport block and value 0 denotes a retransmission.
8. Redundancy version (2 bits), which informs the UE of the HARQ redundancy version for HARQ soft combining.
9. TPC command for PUCCH (2 bits), which informs the UE of the power control information for the uplink.
10. Downlink Assignment Index (2 bits), which informs the UE of the HARQ ACK configuration in TDD mode.
11. SRS request (0 or 1 bit), which triggers aperiodic sounding requests.

DCI formats defined for a particular LTE release can be updated in newer releases. The updated/additional fields in the DCI formats would be understood only by the newer UEs. For example, the field *carrier indicator* was introduced for the cross-carrier scheduling in Carrier Aggregation (CA) in Release 10, where a UE receives or transmits on multiple Component Carriers (CCs). In the absence of cross-carrier scheduling, the downlink scheduling assignment carried on PDCCH is valid for the CC on which it is transmitted, which is similar to legacy carriers. If cross-carrier scheduling is enabled, the PDCCH which carries the DCI information for a particular CC may not be transmitted on the same CC. In this case, the carrier indicator in the DCI provides the index with the CC where the DCI should be applied. Note that the index of the primary CC is always set to 0; even if the eNodeB and UE have a different understanding of the CC index during the period of the RRC reconfiguration, at least the transmission on the primary CC can be scheduled.

One of the major functions of the DCI is to convey the information of the RB assignment for the PDSCH or PUSCH. The signaling design of the RB assignment is to find a good compromise between flexibility and signaling overhead, which is discussed in the next section.

3.5.6.1 PDCCH Transmission and Reception

One PDCCH carries only one DCI for a UE (e.g. the associated PDSCH is used for unicast transmission) or a group of UEs (e.g. the associated PDSCH is used for broadcasting the SIBs). Each DCI message is attached with a CRC, which is then scrambled with a Radio Network Temporary Identifier (RNTI) according to the usage of the PDCCH. If the PDCCH is for a specific UE, the CRC will be scrambled with a unique UE identifier, the Cell-RNTI (C-RNTI). If the DCI is associated with the PDSCH which contains the system information, the CRC will be scrambled with the System Information-RNTI (SI-RNTI). In the case where the corresponding PDSCH contains the paging information, the Paging-RNTI (P-RNTI) will be used for the CRC scrambling.

In general, several PDCCH can be transmitted simultaneously in a subframe. Each PDCCH can occupy one or more Control Channel Elements (CCEs), each of which consists of 9

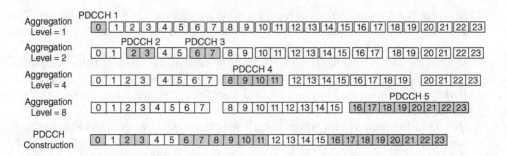

Figure 3.14   Aggregation level and PDCCH construction

REGs. The number of CCEs configured for a particular PDCCH is defined as the aggregation level. The aggregation level for a PDCCH can be configured as 1, 2, 4, and 8, depending on the DCI formats and channel conditions. In Figure 3.14 we provide an example of the PDCCHs with different aggregation levels, where the available CCEs in the control region are indexed from 0 and upwards. The number of available CCEs for PDCCHs is determined by the size of the control channel region, the system bandwidth, the number of antenna ports, and the size of the PHICH.

In LTE, a UE does not know the exact CCE locations of the PDCCHs intended for it. However, the UE knows of a set of CCE locations where it may find its PDCCH. This set of locations is called the search space. A search space is therefore a set of CCEs on a given aggregation level, which the UE is supposed to attempt to decode. Since the UE does not know which DCI format is being carried in the PDCCH, it will perform blind decoding over its search space by checking all possible PDCCH locations and DCI formats. This involves multiple blind decodes and is a task of high complexity. To reduce the complexity, a separate Common Search Space (CSS) and a UE-specific Search Space (USS) are defined in LTE. An example of CSS and USS is given in Figure 3.15. The CSS, as the name implies, is the common search space for all UEs to monitor. The USS is configured for each UE individually. In general, the CSS is used to transmit various system information and paging messages, while the USS is used for unicast transmissions.

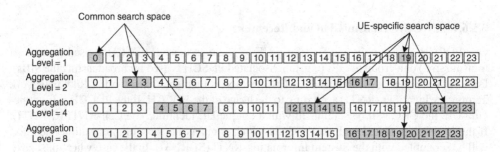

Figure 3.15   Common search space and UE-specific search space

Table 3.5 DCI formats

DCI format	Description	Use case
0	Schedule PUSCH grants	TMs 1–10
1	Schedule PDSCH with a single codeword	TMs 1, 2, and 7
1A	Schedule PDSCH with a compact format	TMs 1–10
1B	Schedule PDSCH for rank-1 transmission	TM 6
1C	Schedule PDSCH with another compact format	MIB and paging
1D	Schedule PDSCH for MU-MIMO	TM 5
2	Schedule PDSCH for closed-loop MIMO	TM 4
2A	Schedule PDSCH for open-loop MIMO	TM 3
2B	Schedule PDSCH for dual-layer beamforming	TM 8
2C	Schedule PDSCH for up to 8-layer multiplexing	TM 9
2D	Schedule PDSCH for CoMP	TM 10
3	Transmit Power Control command (2 bits)	Uplink power control
3A	Transmit Power Control command (1 bit)	Uplink power control
4	PUSCH grants for up to 4-layer multiplexing	TMs 1–10

3.5.6.2 Downlink Control Information (DCI) Formats

LTE specifies a total of 14 DCI formats as shown in Table 3.5. Format 2B was added to support PDSCH TM 8 in Release 9. Formats 2C and 4 were added to support PDSCH TM 9 and PUSCH TM 2, respectively, in Release 10. In PUSCH TM 2, a UE can transmit up to 2 transport blocks per subframe, which can be mapped to 2 or 4 spatial layers. Format 2D was added to support PDSCH TM10 in Release 11. The content of each DCI format is summarized as follows.

Format 0
DCI format 0 is used for the scheduling of PUSCH. The following information is transmitted:

- Carrier Indicator Field (0 or 3 bits);
- Resource allocation type (1 bit);
- Flag for Format 0/1A differentiation (1 bit);
- Frequency hopping flag (1 bit);
- Modulation and coding scheme and redundancy version (5 bits);
- New Data Indicator (1 bit);
- TPC command for scheduled PUSCH (2 bits);
- Cyclic shift for DMRS and Orthogonal Cover Code (OCC) index (3 bits);
- CSI request (1 or 2 bits);
- SRS request (0 or 1 bit);
- UL index (2 bits); and
- Downlink Assignment Index (2 bits).

Format 1

DCI format 1 is used for TMs 1, 2, and 7. The following information is transmitted:

- carrier indicator field (0 or 3 bits);
- resource allocation type (1 bit);
- resource block assignment;
- modulation and coding scheme (5 bits);
- HARQ process number (3 bits for FDD and 4 bits for TDD);
- new data indicator (1 bit);
- redundancy version (2 bits);
- TPC command for PUCCH (2 bits); and
- downlink assignment index (2 bits).

Format 1A

DCI format 1A is used for any TM. The following information is transmitted:

- carrier indicator field (0 or 3 bits);
- flag for Format 0/1A differentiation (1 bit);
- localized/distributed VRB assignment flag (1 bit);
- resource block assignment;
- modulation and coding scheme (5 bits);
- HARQ process number (3 bits for FDD and 4 bits for TDD);
- new data indicator (1 bit);
- redundancy version (2 bits);
- TPC command for PUCCH (2 bits);
- downlink assignment index (2 bits); and
- SRS request (0 or 1 bit).

Format 1B

DCI format 1B is used for TM6. The following information is transmitted:

- carrier indicator field (0 or 3 bits);
- localized/distributed VRB assignment flag (1 bit);
- resource block assignment;
- modulation and coding scheme (5 bits);
- HARQ process number (3 bits for FDD and 4 bits for TDD);
- new data indicator (1 bit);
- redundancy version (2 bits); ·
- TPC command for PUCCH (2 bits);
- downlink assignment index (2 bits);
- TPMI information for precoding; and
- PMI confirmation for precoding (1 bit).

Format 1C

DCI format 1C is used for signaling some system information and paging messages. The following information is transmitted:

- resource block assignment; and
- modulation and coding scheme (5 bits).

Format 1D
DCI format 1D is used for TM5. The following information is transmitted:

- carrier indicator field (0 or 3 bits);
- localized/distributed VRB assignment flag (1 bit);
- resource block assignment;
- modulation and coding scheme (5 bits);
- HARQ process number (3 bits for FDD and 4 bits for TDD);
- new data indicator (1 bit);
- redundancy version (2 bits);
- TPC command for PUCCH (2 bits);
- downlink assignment index (2 bits);
- TPMI information for precoding; and
- downlink power offset (1 bit).

Format 2
DCI format 2 is used for TM4. The following information is transmitted:

- carrier indicator field (0 or 3 bits);
- resource allocation type (1 bit);
- resource block assignment;
- TPC command for PUCCH (2 bits);
- downlink assignment index (2 bits);
- HARQ process number (3 bits for FDD and 4 bits for TDD);
- modulation and coding scheme (5 bits);
- new data indicator (1 bit);
- redundancy version (2 bits); and
- precoding information.

Format 2A
DCI format 2A is used for TM3. The following information is transmitted:

- carrier indicator field (0 or 3 bits);
- resource allocation type (1 bit);
- resource block assignment;
- TPC command for PUCCH (2 bits);
- downlink assignment index (2 bits);
- HARQ process number (3 bits for FDD and 4 bits for TDD);
- transport block to codeword swap flag (1 bit);
- modulation and coding scheme (5 bits);
- new data indicator (1 bit);
- redundancy version (2 bits); and
- precoding information.

Format 2B

DCI format 2B is used for TM8. The following information is transmitted:

- carrier indicator field (0 or 3 bits);
- resource allocation type (1 bit);
- resource block assignment;
- TPC command for PUCCH (2 bits);
- downlink assignment index (2 bits);
- HARQ process number (3 bits for FDD and 4 bits for TDD);
- modulation and coding scheme (5 bits);
- new data indicator (1 bit);
- redundancy version (2 bits);
- scrambling identity (1 bit); and
- SRS request (0 or 1 bit).

Format 2C

DCI format 2C is used for TM9. The following information is transmitted:

- carrier indicator field (0 or 3 bits);
- resource allocation type (1 bit);
- resource block assignment;
- TPC command for PUCCH (2 bits);
- downlink assignment index (2 bits);
- HARQ process number (3 bits for FDD and 4 bits for TDD);
- antenna port(s), scrambling identity and number of layers (3 bits);
- modulation and coding scheme (5 bits);
- new data indicator (1 bit);
- redundancy version (2 bits); and
- SRS request (0 or 1 bit).

Format 3

DCI format 3 is used for the transmission of TPC commands for PUCCH and PUSCH with 2-bit power adjustments.

Format 3A

DCI format 3A is used for the transmission of TPC commands for PUCCH and PUSCH with single bit power adjustments.

Format 4

DCI format 4 is used for the scheduling of PUSCH when PUSCH TM2 is configured. The following information is transmitted:

- carrier indicator field (0 or 3 bits);
- resource allocation type (1 bit);
- resource block assignment;
- TPC command for scheduled PUSCH (2 bits);

- cyclic shift for DMRS and OCC index (3 bits);
- modulation and coding scheme and redundancy version (5 bits);
- new data indicator (1 bit);
- CSI request (1 or 2 bits);
- SRS request (0 or 1 bit);
- UL index (2 bits); and
- downlink assignment index (2 bits).

3.6 Uplink Physical Channels

The uplink physical channels carry uplink data transmission from the UE to the eNodeB. They also carry important control information related to uplink synchronization and downlink channel quality feedback. The initial uplink synchronization and scheduling request is handled by PRACH and the rest falls under the domain of PUSCH and PUCCH.

To understand the details of PUSCH and PUCCH let us first review the different types of information that a UE has to transmit in the uplink, listed as follows.

1. Data transmission.
2. ACK/NACK information about a downlink transmission from the eNodeB.
3. Downlink channel quality indicators (CQI/PMI/RI) that the UE has measured and has to report to the eNodeB.
4. If a UE has data to transmit in the uplink, it has to request the eNodeB to schedule it for uplink transmission. This is transmitted by a message called a Scheduling Request (SR).
5. If the eNodeB responds favorably to the SR, then the UE has to provide further details to the uplink scheduler in the eNodeB. These details are the Buffer Status Report (BSR) which indicates how much data the UE has for transmission and the Power Headroom Report (PHR) which indicates status of the UE transmit power. The scheduler needs these items of information to decide how many RBs to allocate to the UE.
6. In addition to these channels, the UE also has to transmit two reference signals (RSs): the sounding reference signal (SRS) and uplink demodulation reference signal (DMRS) that are used for for uplink channel estimation and demodulation, respectively. Note that the corresponding downlink RSs are CRS/CSI-RS and downlink DMRS, respectively.

Some of these items of information are carried on the PUSCH (such as data, ACK/NACK, CQI/PMI/RI, and BSR/PHR) and some in PUCCH (ACK/NACK, CQI/PMI/RI, and SR). SRS and DMRS are also transmitted along with PUSCH and PUCCH. There are certain rules that govern simultaneous transmissions of PUSCH, PUCCH, and the DMRS/SRS from the same UE in a given subframe. For example the PUSCH and PUCCH are never transmitted simultaneously by a LTE Rel-8 UE in the same subframe.

3.6.1 Physical Uplink Shared Channel (PUSCH)

Similar to the PDSCH, the PUSCH is a shared data channel used by the UEs. There are some differences between the two. When the eNodeB schedules a downlink transmission to a UE in a subframe, the PDSCH contains the data and the PDCCH contains the DCI with the associated control information. Data and associated control are therefore transmitted in the same subframe.

For uplink data transmission from the UE, the transmission parameters are still decided by the eNodeB scheduler. The associated control information for PUSCH transmission is also carried in the PDCCH. In the previous sections we discussed DCI Formats 0 and 4 which carry this control information.

Data transmission in uplink and the associated control channel carrying the transmission parameters do not occur in the same subframe. When the eNodeB transmits an *uplink grant* in subframe N (using DCI formats 0 or 4) for a UE, the UE sends the data in PUSCH in subframe $N + 4$.

The PUSCH also carries control information relevant for downlink transmission. Recall that the UE measures the downlink channel using CRS or CSI-RS to evaluate the CQI and PMI/RI that should be used for the PDSCH transmission. The UE transmits detailed CQI/PMI/RI information using the PUSCH. Note that the UE can also use PUCCH to transmit CQI/PMI/RI. The difference is that PUCCH transmits periodic CQI and PMI/RI information usually of a coarser granularity (e.g. sub-band PMI is not transmitted using PUCCH). This gives the eNodeB some idea of the corresponding downlink channels. If the eNodeB desires more detailed information about the quality of a UE downlink channel it can request the UE to calculate and transmit this information using the PUSCH. CQI and PMI/RI reporting using PUSCH is therefore aperiodic and is said to occur due to the eNodeB *triggering*. This triggering is also carried out using DCI Formats 0 or 4. PUSCH also carries other control information relevant to the downlink transmission, such as ACK/NACK messages.

PUSCH can carry only UE uplink data, only UE downlink control information, or both. In the latter case, they are first multiplexed before the DFT spreading in the uplink.

The signal processing to generate a PUSCH signal is similar to that of the PDSCH. The RBs allocated to a UE for PUSCH transmission are contiguous. Note that this is in contrast to the UE downlink PDSCH transmission where the UE could have been allocated discontiguous bands. This is to maintain low peak-to-average power ratio (PAPR) which is very important in uplink. PUSCH transmission can employ frequency hopping to exploit the frequency diversity gain. Both intra- and inter-subframe hopping is supported in LTE. Whether the UE performs frequency hopping or not is determined by the eNodeB and the UE is notified via the corresponding DCI Formats 0 or 4. The PUSCH transmission carries DMRS in the fourth symbol for all RBs and this is used for uplink demodulation.

3.6.2 Physical Uplink Control Channel (PUCCH)

The PUCCH is used to transmit control information in the uplink similar to PDCCH in the downlink. Unlike the PDCCH, the PUCCH transmission is scheduled for two entire slots that lie on the edges of the total system bandwidth, as shown in Figure 3.16. The PUCCH is always designed to exploit frequency diversity. The two slots are not in the same RB but lie in opposite edges of the total system bandwidth.

Allocating PUCCH resources at the edge of the system bandwidth also allows the PUSCH to be selected from a contiguous block of RBs from the center of the resource grid. The PUCCH region therefore acts as a guard band between PUSCH regions of adjacent carriers.

For the uplink transmission, a general design principle is to ensure that all the RBs in which a UE transmits (PUCCH or PUSCH) should be contiguous. This is needed for maintaining a

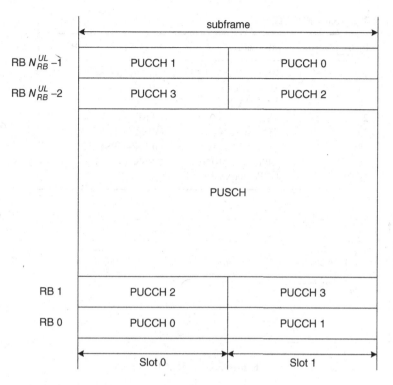

Figure 3.16 PUCCH region

low PAPR after the SC-FDMA modulation in uplink. If the UE is allocated grants for only one of either PUCCH or PUSCH, it transmits in contiguous RBs as the PUCCH is scheduled in edge RBs and the PUSCH in center. The next question that arises is what happens if a UE is scheduled to simultaneously (in the same subframe) transmit PUSCH and PUCCH? In this case, the UE might have to transmit in discontiguous RBs leading to some loss of low-PAPR property. However, this would add more scheduling flexibility and better resource utilization.

The simultaneous transmission of PUSCH and PUCCH is not allowed in LTE Rel-8 but the possibility was introduced in later releases. For example, simultaneous transmission of PUSCH and PUCCH from the same UE is supported in Rel-10 if enabled by higher layers.

The information transmitted by the UE in PUCCH is organized into several PUCCH formats as shown in Table 3.6. This is analogous to the DCI formats in the downlink that were introduced earlier. Different PUCCH formats are used to carry different types of information. The information bits are channel coded, scrambled, and cyclic shifted by a length of 12 base sequence before DFT precoding. Use of the cyclic shift sequence allows up to 12 different UEs to be orthogonally multiplexed in the same PUCCH resource. The DMRS signal is multiplexed in the resulting PUCCH symbols before the IFFT operation and this is used for demodulation at the eNodeB.

PUCCH format 3 was introduced for Rel-10 carrier aggregation (CA) for transmitting multiple ACK/NACKs from multiple component carriers. It follows a different generation procedure than the other PUCCH formats.

Table 3.6 PUCCH formats

PUCCH format	Information	Modulation	Bits/subframe
1	SR	On-Off Keying	N/A
1a	ACK/NACK	BPSK	1
1b	ACK/NACK	QPSK	2
2	CSI	QPSK	20
2a	CSI+ACK/NACK	QPSK+BPSK	21
2b	CSI+ACK/NACK	QPSK+QPSK	22
3	Multiple ACK/NACK + SR (optional)	QPSK	48

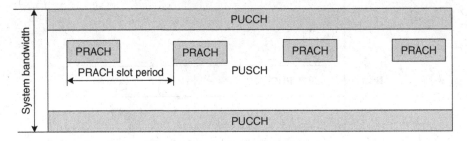

Figure 3.17   PRACH region

3.6.3 Physical Random Access Channel (PRACH)

The PRACH enables initial access and uplink synchronization of the UE. A RACH procedure takes place when the UE is switching on for the first time or is recovering from a radio link failure (RLF) or in any other case when it is switching from a RRC_IDLE to a RRC_CONNECTED state. Even when the UE is in RRC_CONNECTED state but not yet uplink synchronized, or is in the process of handover to another cell, RACH will be used for uplink synchronization. Recall that prior to attempting uplink synchronization, the UE has already achieved downlink synchronization by decoding the PSS and SSS.

The subframes where UE transmits RACH are shown in Figure 3.17, which is time and frequency multiplexed with the PUSCH and PUCCH. The location of these subframes are indicated to the UE via SIB-2 which the UE has already decoded during the downlink synchronization process.

The PRACH procedure consists of the following basic steps.

1. **RACH Preamble Transmission:** The UE chooses a preamble from a set of RACH preambles. The number of possible preambles is 64. Out of these, some are for *contention-based* access which means that a UE picks any one of them randomly. Two different UEs may pick the same preamble, leading to subsequent contention. The other preambles are used for *contention-free* access which means that the eNodeB informs the UE via RRC signaling which preamble it should use. Different UEs therefore have different preambles and there is no contention. This is used for RACH transmissions during handover when the UE is already in RRC_CONNECTED state.

PRACH is transmitted by special OFDM modulation with 1.25 kHz subcarrier spacing and 800 μs symbol duration to increase reliability. The transmit power for initial RACH transmission is based on open loop estimation of the path-loss.

2. **Random Access Response:** Upon successful reception of the UE preamble, the eNodeB transmits the Random Access Response (RAR). The RAR is transmitted on the PDSCH and is scrambled by the RA-RNTI which is known to the UE. It contains the identity of the detected preamble, a timing alignment message for uplink synchronization which is the assignment of a temporary C-RNTI to the UE. It always contains a small PUSCH grant for subsequent transmission from the UE.

   If multiple UEs had transmitted the same preamble, the eNodeB may still be able to decode it. If so, then each one of these UEs would receive and be able to decode the subsequent RAR. If the UE does not receive a RAR within a configured time window, it goes back to step (1) and may ramp up the transmit power.

3. **RRC Connection Request:** The UE uses the PUSCH grant to transmit an RRC connection request. The UE also transmits information about its identity. This could be the SAE Temporary Mobile Subscriber Identity (S-TMSI) in the case of initial synchronization or the actual C-RNTI of the UE in the case of an RRC_CONNECTED UE. It is scrambled by the temporary C-RNTI obtained from the previous step.

   If multiple UEs received the RAR message in the previous step, all of them send the RRC connection request which will again collide at the eNodeB. Since the messages are now different (as opposed to step (1) where each UE had transmitted the same preamble) there would be interference at the eNodeB. The eNodeB may not be able to decode message from any UE in which case all the UEs go back to step (1). One UE may be successfully decoded however, allowing progression to the next step.

4. **Contention Resolution:** The eNodeB sends the contention resolution message addressed to the C-RNTI or S-TMSI of the UE whose message was successfully decoded in the previous step. All UEs which transmitted in the last step decode the message and only the UE which detects its own identity responds with a HARQ. The other UEs deduce that there was a collision and return to step (1). The RACH procedure is now complete.

References

[1] 3GPP (2012) Evolved Universal Terrestrial Radio access (E-UTRA) and Evolved Universal Terrestrial Radio Access (E-UTRAN), overall description 3GPP TS 36.300 v11.2.0. Third Generation Partnership Project, Technical Report.

[2] Hitachi (2011) R1-112587, System Design Considerations for CoMP and eICIC. Technical Report, August 2011.

[3] 3GPP (2012) Evolved Universal Terrestrial Radio access (E-UTRA), Multiplexing and channel coding 3GPP TS 36.212 v11.0.0. Third Generation Partnership Project, Technical Report.

[4] 3GPP (2012) Evolved Universal Terrestrial Radio access (E-UTRA), Physical layer procedures 3GPP TS 36.213 v11.0.0. Third Generation Partnership Project, Technical Report.

[5] Sesia, S., Toufik, I., and Baker, M. (2011) *LTE: The UMTS Long Term Evolution: From Theory to Practice*. United Kingdom: John Wiley & Sons Ltd.

4

Physical Layer Signal Processing in LTE

4.1 Introduction

This chapter focuses on the LTE physical layer procedural aspects that are necessary before any data can be sent across the air interface between the eNodeB and the UE. The LTE system defines the following series of major physical layer procedures to establish the communication between the two entities.

1. Synchronization and cell search, by which a UE acquires the cell ID, slot and radio frame timing synchronization, and various other critical system parameters that are necessary before a UE can associate with an eNodeB.
2. Reference signal detection, which is required for the channel measurements as well as the demodulation of various control and data channels by both the UE and the eNodeB.
3. Channel estimation and feedback to the eNodeB, which involves a UE estimating the underlying radio channel, reporting quantized channel quality information to the eNodeB to enable it to dynamically adapt ensuing data transmissions to match the channel capacity.
4. Scheduling and resource allocation, by which the eNodeB allocates radio resources to a selected subset of UEs by striking a reasonable balance between the desired quality of service (QoS) for various UEs and the optimization of overall system throughput.

4.2 Downlink Synchronization Signals

When a UE powers on it has no knowledge of the system parameters, including synchronization. The first step is therefore for the UE to perform cell synchronization by acquiring downlink synchronization signals. In a LTE network, downlink synchronization signals are transmitted twice in every radio frame to enable a UE to acquire time and frequency synchronization. Two downlink synchronization signals are defined in LTE: Primary Synchronization Signal (PSS) and Secondary Synchronization Signal (SSS). The PSS acquisition is the first step in the cell synchronization process and is followed by the acquisition of the SSS. Upon successful acquisition of the PSS and SSS, a UE synchronizes to OFDM symbol, slot,

Heterogeneous Networks in LTE-Advanced, First Edition. Joydeep Acharya, Long Gao and Sudhanshu Gaur.
© 2014 John Wiley & Sons, Ltd. Published 2014 by John Wiley & Sons, Ltd.
Companion Wesite: www.ltehetnet.com.

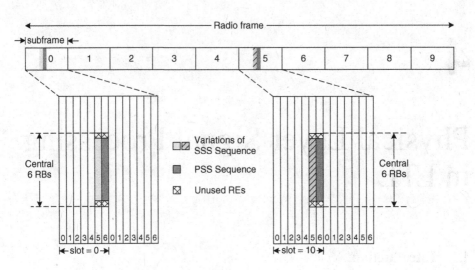

Figure 4.1   Location of PSS and SSS in a FDD radio frame with normal CP

subframe, and radio frame boundaries. The cell acquisition process also enables a UE to determine the physical cell identity (PCI) of the concerned cell. The UE then proceeds to perform PBCH acquisition and other related tasks.

Figure 4.1 shows the location of the PSS and SSS signals for a Frequency Division Duplex (FDD) radio frame. These synchronization signals are transmitted twice within a radio frame and are located in the central 6 PRBs around the DC subcarrier. This simplifies the cell search procedure as the location of these synchronization signals remains the same regardless of the system bandwidth, which is unknown to a UE *a priori*.

The time domain location of the PSS and SSS varies with frame structure (FDD or Time Division Duplex, TDD). In an FDD system the PSS and SSS are placed adjacently within one subframe, while in a TDD system they are placed in different subframes separated by two OFDM symbols. Table 4.1 shows the location of these synchronization signals for TDD and FDD radio frames in the case of normal and extended cyclic prefixes. Both the PSS and SSS are transmitted via the same antenna port to allow for coherent detection of the SSS relative to the PSS. However, different antenna ports may be utilized across the subframes to allow for time-switched antenna diversity gain.

As shown in Figure 4.1, the PSS and SSS sequences do not occupy all 72 subcarriers spanning the central 6 PRBs. Both the PSS and SSS are of equal length and mapped to 31 subcarriers on either side of the DC carrier as shown in Figure 4.2. This helps a UE to detect the PSS and SSS, utilizing a lower complexity FFT of size 64. Additionally, the resulting 5 unused subcarriers on either side of the PSS and SSS provide the additional benefit of guard band protection from other PDSCH transmissions.

4.2.1   Primary Synchronization Signal

In a FDD cell, the PSS is transmitted on the first slots of subframes 0 and 5 in every radio frame in the case of normal cyclic prefix. Although the PSS sequence carries only sector information

Table 4.1 PSS and SSS sequence locations in the time domain

		PSS		SSS	
Duplexing mode	Cyclic prefix	(slots)	(symbol)	(slots)	(symbol)
FDD	Normal	0, 10	6	0, 10	5
	Extended	0, 10	5	0, 10	4
TDD	Normal	2, 12	2	1, 11	6
	Extended	2, 12	2	1, 11	5

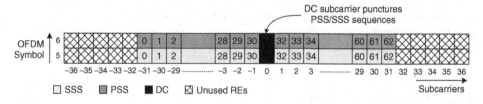

Figure 4.2 Mapping of PSS and SSS sequences in a FDD cell with normal CP

(N_{ID}^2), its location acquisition also enables a UE to determine OFDM symbol, slot, and subframe boundaries. Since PSS sequences are repeated every 5 ms, its acquisition also enables a UE to determine half-frame boundaries. However, PSS acquisition does not enable a UE to determine radio frame boundary, cyclic prefix, or complete physical cell identity (PCI), N_{ID}^{cell}, of the concerned cell.

The PSS sequences are derived from the frequency domain Zadoff Chu (ZC) sequences of length 63. In LTE, the complex value ZC sequence used $d_u(n)$, where u is the ZC parameter, is given by [1]:

$$d_u(n) = \exp\left[-j\frac{\pi un(n+1)}{63}\right], \qquad n = 0, 1, \ldots, 30, 32, \ldots, 62. \qquad (4.1)$$

It may be noted that, since the DC carrier is not used in LTE, the PSS sequence is punctured in the middle at $n = 31$ as shown in Figure 4.2. In LTE 3 PSS sequences are defined to convey the physical cell identity of the cell. These cell identities are mapped to 3 distinct roots of ZC sequence u as shown in Table 4.2.

Table 4.2 ZC root indices for PSS generation

Physical layer identity $(N_{ID}^{(2)})$	ZC root index (u)
0	25
1	29
2	34

(Source: 3GPP 2012 [1]. Reproduced with permission of 3GPP.)

These 3 ZC sequences were chosen for their good periodic autocorrelation and cross-correlation properties. In particular, these sequences have a low-frequency-offset sensitivity allowing for a robust PSS detection during the initial synchronization process. Due to this good autocorrelation property and low-frequency-offset sensitivity, the PSS can be easily detected with a frequency offset up to ± 7.5 kHz. Additionally, the sequences corresponding to the roots 29 and 34 are the complex conjugate of each other and can be detected using a single correlator (thus simplifying UE complexity).

Since the PSS acquisition is the first step in the cell synchronization procedure, only non-coherent detection is possible. In fact, the PSS detection is the only necessary non-coherent detection required in LTE signaling. Since the channel phase information is unavailable, a non-coherent maximum likelihood (ML) detector is utilized for the PSS detection. Additionally, the PSS detection also allows a UE to perform coherent detection for the SSS, assuming the channel remains constant for the duration of these two synchronization signals.

4.2.2   Secondary Synchronization Signal

Acquisition of the SSS is the next step after the PSS timing detection in the cell synchronization process. Different from the PSS transmission, the SSS sequence transmission varies in a specific manner for the first and second occurrences within a radio frame. These two variations of the SSS sequence are needed to enable a UE to acquire a radio frame boundary and thus complete the cell synchronization process. The SSS transmission can occur at 4 different locations with respect to the PSS transmission as shown in Table 4.1. The relative location of SSS occurrences within the radio frame enables a UE to determine the cyclic prefix as well as duplexing mode (FDD or TDD) used in the cell. Apart from assisting in cell synchronization, the SSS transmission also informs the UE of the cell identity group (N_{ID}^1) which ranges from 0 to 167. Thus, a successful acquisition of the SSS together with the PSS enables a UE to completely determine the Physical Cell Identity (PCI) (N_{ID}^{cell}) of the concerned cell, which varies from 0 to 503, as:

$$N_{ID}^{cell} = 3N_{ID}^1 + N_{ID}^2. \tag{4.2}$$

An eNodeB generates the SSS sequence from 3 maximum-length base sequences (s, c, and z) also known as M-sequences. These M-sequences are pseudo-random binary sequences of length 31 constructed by cycling through all possible states of a shift register of length 5. Figure 4.3 shows the SSS sequence generation for transmission in the 1st slot of a FDD radio frame. The M-sequences $s_0^{(m_0)}$, $s_1^{(m_1)}$, and $z_1^{(m_0)}$ are derived from the respective base sequences by applying unique cyclic shifts which are dependent upon the cell identity group (N_{ID}^1). Similarly, the scrambling sequences c_0 and c_1 are cyclically shifted versions of the base sequence c where cyclic shifts are based on the physical cell identity (N_{ID}^2). The sequences $s_0^{(m_0)}$, and $s_1^{(m_1)}$ are scrambled by scrambling codes c_0 and c_1. The second sequence goes through additional scrambling by another binary sequence $z_1^{(m_0)}$. The resulting two sequences are then interlaced, Binary Phase Shift Keying (BPSK) modulated and then mapped to 31 subcarriers on either side of the DC subcarrier. Since one of the main functions of the SSS is to enable radio frame boundary detection, two different variations of the SSS sequence are transmitted in a radio frame. The variation is achieved by switching $s_0^{(m0)}$ and $s_1^{(m1)}$ for alternative transmissions. The SSS is mapped to the same set of subcarriers as the PSS sequence as shown in Figure 4.2. A UE

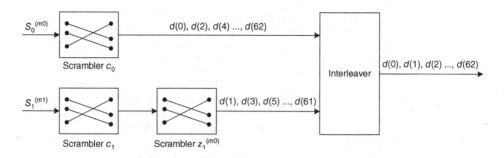

Figure 4.3   SSS sequence generation in a FDD radio frame for 1st slot

observing these specific variations in the SSS sequence can correctly determine the radio frame boundary. Once a UE acquires the SSS it can determine the cell identity group based on the induced cyclic shift to one of the M-sequences relative to the base sequence.

Unlike the non-coherent detection used for the PSS acquisition, coherent detection as well as non-coherent detection techniques can be used for the SSS sequence. Since the SSS detection follows the PSS detection, the channel can be estimated based on the PSS sequence assuming the channel does not vary by much for the duration of the PSS and SSS transmissions. A coherent ML detector could therefore be employed for SSS detection. However, coherent detection is not always desirable if the UE happens to be on the cell edge of two or more cells using the same PSS sequences, as this leads to incorrect channel estimation based on the PSS. Such a scenario is often observed in a heterogeneous setting where picocell UEs can hear multiple cells, with some using the same physical cell identity, leading to the same PSS transmission. In such cases, non-coherent detection for the SSS is expected to perform better than coherent detection, assuming the coherence bandwidth of the channel is greater than the bandwidth spanned by the 6 PRBs which carry the SSS.

4.3   Reference Signals

In LTE, various reference signals are utilized in the downlink and the uplink for channel measurements, enabling different MIMO transmission modes and demodulation of control and data channels. For the downlink channel, each reference signal pattern is associated with an antenna port at the eNodeB. In LTE, an antenna port is defined as a combination of a subset of eNodeB antennas that are perceived as a single transmit antenna by the UE. Thus, each antenna port defines a unique channel that carries the corresponding RS and the control or data channel associated with the RS. Table 4.3 shows the mapping between various antenna ports and the associated RSs. Note that when an RS is transmitted corresponding to a given antenna port, the other antenna ports are muted to allow for channel estimation for that antenna port.

4.3.1   Downlink Reference Signals

The various downlink reference signals such as CRS, CSI-RS, IMR, and DMRS are described in the following sections.

Table 4.3 Antenna port mapping for different RSs

Antenna port	Configured RS
0–3	CRS
4	MBSFN RS
5	DMRS for single-layer beamforming
6	Positioning RS
7–8	DMRS for dual-layer beamforming
9–14	DMRS for multi-layer beamforming
15–22	Channel state information (CSI) RS

4.3.1.1 Cell-Specific Reference Signals

The cell-specific reference signal (CRS) is the first reference signal that a UE attempts to detect after the acquisition of the downlink synchronization signals. It enables the coherent detection of the PBCH as well as several other downlink control channels and PDSCH for transmission modes 1–5. Additionally, CRS is used for Radio Resource Management (RRM) measurements such as Reference Signal Received Power (RSRP) and Reference Signal Received Quality (RSRQ) required for scheduling and handovers.

The CRS symbols are derived from a 31-length gold sequence which is generated with the seed based on the slot number, OFDM symbol number, cyclic prefix type, and cell identity. The generated sequence is therefore cell-specific and enables a UE to differentiate the CRS from different cells. While the sequence itself is $2^{31} - 1$ bits in length, the number of bits from the sequence selected for transmission is based on the largest channel bandwidth, which is currently set as $N_{RB}^{max} = 20$ MHz. The bit sequence for CRS generation is extracted from the center of the truncated sequence to ensure that the same CRS symbols are present around the PBCH regardless of the channel bandwidth. Once the UE decodes the PBCH and determines the system bandwidth, it can generate the complete CRS sequence for a given OFDM symbol by extracting $4N_{RB}$ bits from the center of the truncated sequence. The CRS symbols are obtained using the QPSK modulation of the $4N_{RB}$ bits resulting in a density of 2 CRS symbols per RB.

In LTE, the CRS symbols are mapped to the REs forming in a uniform reference symbol grid across all the RBs spanning the system bandwidth. This configuration is known to be optimal for OFDM-based system. Additionally, LTE defines CRS for up to 4 antenna ports with unique CRS to RE mapping for each antenna port. Figures 4.4 and 4.5 show the CRS mapping for antenna ports 0 and 1 respectively. The REs used for CRS symbols on one antenna port are reserved and are not used on other antenna ports to minimize CRS interference. For higher-numbered antenna ports, increasing CRS density linearly can result in significant overheads, nullifying any throughput gain from better channel estimation. For this reason, reduced-density CRS are utilized for antenna ports 2 and 3 as shown in Figure 4.6.

There is also a frequency shift (N_{ID}^{cell} mod 6) that applies to the CRS mapping shown in Figures 4.4, 4.5, and 4.6 to avoid CRS collision due to PDSCH transmissions from the neighboring cells. This is an important feature as otherwise interference from CRS REs could be a severe issue because of the provision of transmission power boosting for CRS REs by up to 6 dB. In densely deployed heterogeneous networks the possible 6 frequency shifts may not be

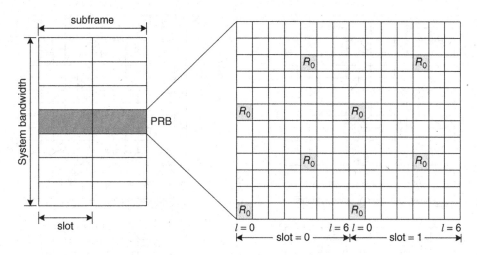

Figure 4.4 Mapping of cell-specific reference symbols (R_0) for 1 antenna port in a subframe with normal CP. (Source: 3GPP 2012 [1]. Reproduced with permission of 3GPP.)

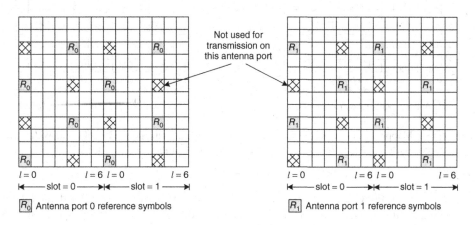

Figure 4.5 Mapping of cell-specific reference symbols for 2 antenna ports in a subframe with normal CP. (Source: 3GPP 2012 [1]. Reproduced with permission of 3GPP.)

sufficient to avoid CRS collisions. This issue was considered in 3GPP and led to the design of advanced CRS-IC receivers in Release 11.

4.3.1.2 Channel State Information Reference Signal

The CRS enables the UE to estimate the CSI for up to 4 antenna ports. In Release 10, multi-antenna transmission was extended to support downlink spatial multiplexing up to 8 layers in Transmission Mode (TM) 9. In order to support the CSI measurement up to 8 antenna ports, Channel State Information Reference Signal (CSI-RS) was introduced.

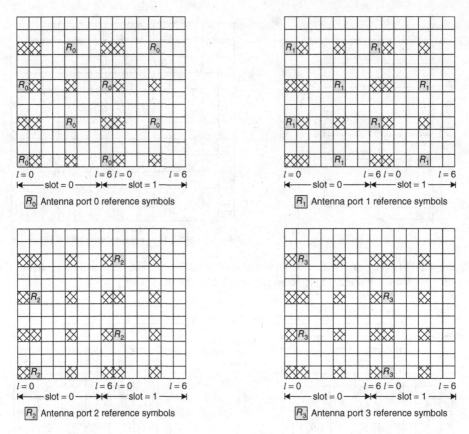

Figure 4.6 Mapping of cell-specific reference symbols for 4 antenna ports in a subframe with normal CP. (Source: 3GPP 2012 [1]. Reproduced with permission of 3GPP.)

Different from the CRS, which could be used for CSI measurement and/or demodulation depending on the transmission mode (TM) (see Section 4.4.1), the CSI-RS is used for CSI measurement only. For example, the CRS is used for both CSI measurement and demodulation in TM4. In TM9, the CSI-RS is used for CSI measurement only while demodulation is based on the Demodulation Reference Signal (DMRS), allowing the CSI-RS to have lower time/frequency density and thus lower system overheads. It is also desirable to have low-density CSI-RS in order to minimize the impact to legacy UEs, since they are unaware that their PDSCH data are punctured by CSI-RS. Considering the trade-off between the CSI estimation accuracy and the performance degradation of legacy UEs, it was decided that the density of CSI-RS is one RE per RB per antenna port.

The CSI-RS patterns specified in Release 10 are shown in Figure 4.7. There are a total of 40 REs that could potentially be used for the CSI-RS transmission, which are grouped into pairs. In a given cell, a subset of these REs is used for the CSI-RS transmission. In the case where two antenna ports are configured, the corresponding two CSI-RS ports use a pair of REs and are multiplexed by applying Orthogonal Cover Code (OCC) to the two REs. There are therefore 20 reuse patterns, i.e. C1–C20 as shown in Figure 4.7, which allow different

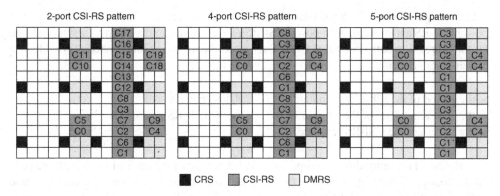

Figure 4.7   CSI-RS patterns for LTE-Advanced. (Source: 3GPP 2012 [1]. Reproduced with permission of 3GPP.)

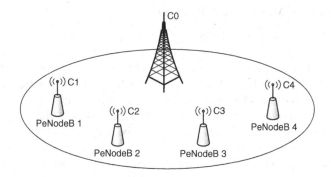

Figure 4.8   An example of CSI-RS resource configuration

cells to utilize different patterns to avoid multiple CSI-RS collisions. An example of allocating different CSI-RS resources to different macro/picocells is given in Figure 4.8. In the case where 4/8 antenna ports are configured, the corresponding CSI-RS ports are multiplexed with OCC on 2/4 pairs of REs; it is therefore possible to have 10/5 non-overlapping CSI-RS patterns as shown in Figure 4.7.

In the time domain, the CSI-RS in different cells can be transmitted with different periods (ranging from 5 ms to 80 ms) and subframe offsets. This can avoid the CSI-RS collision from different cells using the same CSI-RS pattern. In the subframe where the CSI-RS is transmitted, it is transmitted in every Resource Block (RB) in the frequency domain such that the UE can perform the CSI measurement across the entire bandwidth.

As mentioned earlier, only a subset of the 40 CSI-RS REs is used to transmit CSI-RS for a given cell. The remaining REs can be used for PDSCH transmission by default. In Release 10, it is also possible for the cell to mute certain CSI-RS REs in order to reduce the interference to the CSI-RS transmissions in its neighboring cell. In the example shown in Figure 4.8, the MeNodeB can perform muting on CSI-RS REs C1–C4 to reduce the interference to the CSI-RS transmissions in the picocells. The muted CSI-RS is defined as zero-power CSI-RS, which follows the same pattern as normal non-zero power CSI-RS but carries nothing.

The configuration of CSI-RS is UE-specific. For a particular UE in Release 10, one non-zero power CSI-RS and multiple zero-power CSI-RS resources transmitted from its serving cell can be configured. For each configured CSI-RS resource, the following information is provided via higher-layer signaling:

- the period of the subframes containing the CSI-RS resource;
- the subframe offset of the subframes containing the CSI-RS resource; and
- the pattern of the CSI-RS resource.

In Release 11, multiple non-zero-power CSI-RS resources transmitted from multiple cells can be configured for a UE in order to facilitate the CoMP operation. More details can be found in Chapters 8 and 9. The PDSCH mapping for the UE is modified to avoid the REs configured for zero-power and non-zero-power CSI-RS resources.

4.3.1.3 Interference Measurement Resource

In Release 11, the zero-power CSI-RS could be used for a UE to measure the interference from its serving cell (for MU-MIMO operation) and/or its neighboring cells (for CoMP operation). In particular, Interference Measurement Resource (IMR) was introduced, which is configured as a 4-port zero-power CSI-RS resource. The IMR allows a UE to measure the interference from its neighboring cells on the configured 4 REs. Even if the interference occurs on partial IMR REs, the UE measures the interference on all of them for signaling simplicity.

Several different ways of configuring IMRs have been proposed [2] and it has been suggested that a large number of IMRs is needed for network operation. In particular, it has been suggested that the required reuse factors for the enhanced interference estimation mechanism should not be lower than the number of current cell IDs.

The number of IMRs is very much dependent on the deployment scenario. In future we may expect to have a very dense collection of nodes in a geographical area. The interference situation may also change dynamically. To increase the efficiency of network operations such as CoMP, a UE may have to measure many different kinds of interference hypotheses. A large number of IMRs would be beneficial for measuring interference in such situations.

Due to the dynamic formation of cells, it may be difficult for operators to configure this large number of IMRs dynamically. This is especially true if nodes are deactivated/activated dynamically. In such a case a large number of pre-defined IMRs would help with network planning. One possible example is to have a pre-decided configuration of IMRs for each cell as a function of the cell IDs. This is a simple solution, but not the only one. This method of configuring IMRs was not adopted for Release 11 due to lack of consensus but may be considered for later releases, especially for dense small-cell deployments.

4.3.1.4 UE-Specific Reference Signals

In addition to cell-specific RSs, a UE can also be configured to receive UE-specific RSs (also known as demodulation reference signals or DMRS) to enable downlink beamforming for PDSCH transmission. Different from CRS, a DMRS sequence is initialized using the UE's identity rather than the cell ID and only spans the RBs allocated to the UE. The DMRS symbols undergo the same precoding that is applied to the PDSCH data symbols. A UE configured with DMRS is therefore required to use it to estimate the precoded channel for demodulating the PDSCH data. Note that for the purposes of Channel Quality Indicator (CQI) and Precoding Matrix Indicator (PMI) feedback, the UE still uses CRS-based channel estimation.

Figure 4.9 shows the arrangement of DMRS on antenna port 5 for single-layer beamforming. Note that while the DMRS arrangement is similar to that of CRS, its density is kept lower to minimize the increase in overheads. LTE Releases 9 and 10 introduced new DMRS antenna ports to support multi-layer beamforming. Figure 4.10 shows the DMRS mapping for up to rank 4 beamforming.

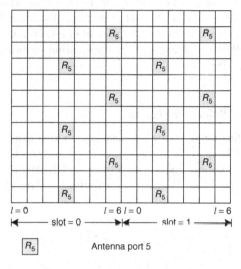

Figure 4.9 LTE Release 8 mapping of UE-specific DMRS symbols for single-layer PDSCH transmission with normal CP

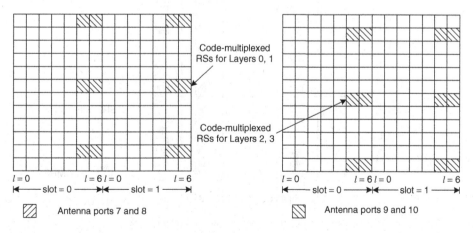

Figure 4.10 LTE Release 10 mapping of UE-specific DMRS symbols for rank 4 PDSCH transmission with normal CP. Different layers are CDM and FDM multiplexed

4.3.2   Uplink Reference Signals

In this section we discuss the various uplink reference signals such as SRS and DMRS.

4.3.2.1   Sounding Reference Signals

The Sounding Reference Signal (SRS) is a wideband uplink transmission by a UE that allows the eNodeB to estimate the uplink channel quality over RBs spanning beyond the UE's allocated PUSCH region. The eNodeB can use this information for various purposes, including frequency selective uplink scheduling and uplink power control. The SRS transmission can be configured by the eNodeB from every 2 ms to 160 ms, thus offering a trade-off between overheads and the quality of uplink channel estimation. As shown in Figure 4.11, the SRS transmission always occurs in the last OFDM symbol of a subframe. All the UEs in a cell are informed by the eNodeB of the impending SRS transmission, ensuring that no collision occurs between the PUSCH and the SRS.

4.3.2.2   Demodulation Reference Signals

To facilitate the demodulation of the uplink data and control channels, a UE embeds DMRS when it transmits PUSCH or PUCCH. Similar to the downlink channel, the DMRS enables coherent detection of the uplink transmissions from the UE. The DMRS spans only the RBs allocated for the PUSCH or PUCCH transmission for the UE. The DMRS occupies the middle

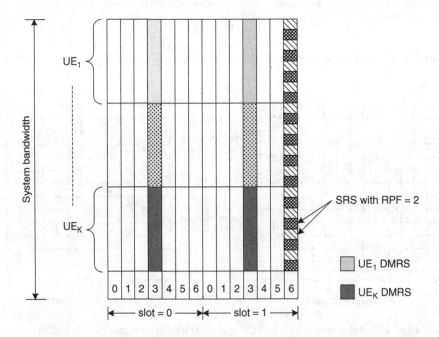

Figure 4.11   Mapping of SRS and DMRS in a PUSCH transmission with normal CP

OFDM symbol in each slot and is transmitted along with the PUSCH data in a TDM fashion as shown in Figure 4.11, where multiple UEs have non-overlapping PUSCH regions. The orthogonal nature of the UE-specfic DMRS sequences makes it possible for multiple UEs to share the same PUSCH region and thus allow MU-MIMO operation in the uplink channel. Note that the density and arrangement of DMRS for PUCCH transmission is different from PUSCH and is uniquely defined for each PUCCH format.

4.4 Channel Estimation and Feedback

In this section we discuss channel estimation and feedback in LTE. Unlike UMTS, which used closed-loop power control to guarantee a fixed rate stream to a UE, LTE enables channel-adaptive variable-rate transmission to a UE. This link adaptation is enabled by channel estimation and feedback from the receiver. In this section we discuss the downlink link adaptation, where the UE estimates the channel and feeds back a quantized version to the eNodeB. The eNodeB receives this information with a finite delay and with possible transmission errors. Based on this information, the eNodeB adapts its transmission strategies to the UE. Possible transmission strategies at the eNodeB that are determined based on UE feedback include:

- choice of transmit precoder that maps the data streams to the antennas before transmission;
- choice of modulation and coding schemes, which give the transmission rate;
- decisions made at the eNodeB scheduler;
- for later LTE releases, UE feedback also determines the performance of advanced signal processing such as base station coordination.

In this section we will discuss the design principles of UE feedback and how this is utilized at the eNodeB.

4.4.1 Basics of Link Adaptation

The basics of transmitter and receiver processing in LTE were covered in Section 1.4. Let us first reproduce the transmission model of a SU-MIMO downlink transmission with transmit precoder $\mathbf{W}$ and receive filter $\mathbf{F}$:

$$\mathbf{r} = \mathbf{F}^H \mathbf{y} = \mathbf{F}^H \mathbf{H} \mathbf{W} \mathbf{s} + \mathbf{F}^H \mathbf{z}. \tag{4.3}$$

The UE estimates $\mathbf{H}$ and has to convey this information to the eNodeB. Practical feedback constraints prohibit transmission of continuous complex-valued channel matrix $\mathbf{H}$. The UE thus feeds back quantized information about $\mathbf{H}$ to the eNodeB.

The first item of information is about the transmit precoder that the eNodeB should use during data transmission. LTE defines a finite set of possible matrices $\mathbf{W}$ that can be used as precoders. This set, which we denote C, is called the codebook. The eNodeB and UE each maintain the codebook. The LTE Release 8 codebook structure is shown in Table 4.5 for a eNodeB with 2 antenna ports. For two antenna ports at the transmitter, two independent data streams at most can be transmitted. The precoders in the codebook are thus either for rank 1 or rank 2 transmission. For more details, including codebooks for a higher number of antenna ports, refer to [1].

Given a codebook C, the UE can select a precoder $\mathbf{W}^*$ by maximizing the channel norm:

$$\mathbf{W}^* = \arg \max_{\mathbf{W} \in C} \|\mathbf{H}\mathbf{W}\| \tag{4.4}$$

The optimization is performed over all possible transmission ranks and all possible precoders per rank. The UE then transmits the optimal transmission rank called the Rank Indicator (RI) and the index of $\mathbf{W}^*$ in C for rank RI. This index is called the Precoding Matrix Indicator (PMI). The pair (PMI, RI) therefore identify a unique entry in the codebook C.

How is the codebook C designed? A precoder should be matched to the eigendirections of the channel correlation matrix. Since the channel fading matrix is random, identifying the precoder structure is not straightforward and has hence attracted much research attention [3, 4]. Most of these works assume a tractable fading model such as Rayleigh, which are not valid for realistic channel models. However 3GPP channels do include correlations introduced due to the uniform linear array (ULA) antenna structure. Recall Example 2.5.6 in Chapter 2, which depicted these correlations for LoS channels. Even in the presence of short-term fading these correlations still exist. Codebook entries can therefore be aligned to match the ULA correlations.

Let us first consider design guidelines for rank 1 precoding vectors. Define the vector $a_P(\theta) = [1, e^{j\theta}, \cdots, e^{j(P-1)\theta}]$. If all N_T transmit antennae at the eNodeB have the same polarization, then vectors of the form $a_{N_T}(\theta)^T$ with uniformly distributed angles θ can be shown to be the eigenvectors of the ULA correlation matrix. For cross-polarized antenna arrays, where half of the antennae have vertical polarization and the other half have horizontal polarization, the corresponding eigenvectors are of the form $[a_{N_T/2}(\theta), \alpha a_{N_T/2}(\theta)]^T$ where α is the phase difference adjustment factor between the two polarizations.

The LTE rank 1 codebook therefore has a mixture of vectors of the form $a_{N_T}(\theta)^T$ and $[a_{N_T/2}(\theta), \alpha a_{N_T/2}(\theta)]^T$. The precoders for higher ranks are obtained from the rank 1 vectors via the 'Householder' generating function to yield precoding matrices with orthogonal columns.

The RI and PMI essentially contain information about the eigenvectors of the channel covariance matrix. The eigenvalue information also has to be reported by the UE. This is proportional to the received SINR per layer after the receiver processing filter operation $\mathbf{F}$. However, the UE does not directly feed back a quantized version of this SINR. LTE defines a set of modulation and coding scheme (MCS) values that the eNodeB uses during actual data transmission. This set of MCS values is available to the UE. The process of adapting the modulation and coding parameters based on UE feedback is called Adaptive Modulation and Coding (AMC). The UE calculates the post-processing SINR and then computes the highest MCS value for which it can receive data with block error rate (BLER) probability not exceeding 10%. This value is called the Channel Quality Indicator (CQI). Table 4.4 shows mapping of CQI indices to different MCS schemes.

The number of CQI values reported is equal to the number of independent data streams (called codewords) at the transmitter. Each codeword is generated by an independent channel encoder and modulator block. The number of codewords is less than or equal to the number of layers of the channel matrix. In LTE a maximum of two codewords is defined but the number of layers can be up to four for LTE Release 8 and up to eight for Release 10, which supports

Table 4.4 CQI table for different MCS schemes

CQI index	Modulation	Code rate × 1024	Efficiency
0	–	–	–
1	QPSK	78	0.1523
2	QPSK	120	0.2344
3	QPSK	193	0.3770
4	QPSK	308	0.6016
5	QPSK	449	0.8770
6	QPSK	602	1.1758
7	16QAM	378	1.4766
8	16QAM	490	1.9141
9	16QAM	616	2.4063
10	64QAM	466	2.7305
11	64QAM	567	3.3223
12	64QAM	666	3.9023
13	64QAM	772	4.5234
14	64QAM	873	5.1152
15	64QAM	948	5.5547

(Source: 3GPP 2012 [1]. Reproduced with permission of 3GPP.)

Table 4.5 LTE Release 8 codebooks for up to two transmit layers

Codebook index (PMI)	One layer (RI = 1)	Two layers (RI = 2)
0	$\frac{1}{\sqrt{2}}\begin{bmatrix}1\\1\end{bmatrix}$	$\frac{1}{\sqrt{2}}\begin{bmatrix}1 & 0\\0 & 1\end{bmatrix}$
1	$\frac{1}{\sqrt{2}}\begin{bmatrix}1\\-1\end{bmatrix}$	$\frac{1}{\sqrt{2}}\begin{bmatrix}1 & 1\\1 & -1\end{bmatrix}$
2	$\frac{1}{\sqrt{2}}\begin{bmatrix}1\\j\end{bmatrix}$	$\frac{1}{\sqrt{2}}\begin{bmatrix}1 & 1\\j & -j\end{bmatrix}$
3	$\frac{1}{\sqrt{2}}\begin{bmatrix}1\\-j\end{bmatrix}$	

(Source: 3GPP 2012 [1]. Reproduced with permission of 3GPP.)

eight antenna systems. LTE defines a mapping between the data symbols of codewords to the data symbols that are mapped to the transmit layers.

The CQI values calculated by the UE depend not only on the channel matrix but also on the receive filter. Two UEs with identical channels may report different CQI values if their receive filters are different. For example, a UE with a MMSE receiver with interference cancellation capabilities could report a higher CQI value than a UE with a traditional MMSE receiver.

Note that the CQI, PMI, and RI values that the UE feeds back are only recommendations to the eNodeB. During actual data transmission, the eNodeB may derive other values of transmit precoder and MCS. The final eNodeB-selected values of precoder and MCS are transmitted to the UE via DCI as explained in Chapter 3. Let us consider an instance of this now, namely MU-MIMO CQI reporting.

Recall the MU-MIMO transmission that was described in Section 1.4. For the two UE cases, the transmission model was given by

$$r_1 = F_1^H y_1 = F_1^H H_1 (W_1 s_1 + W_2 s_2) + F_1^H z_1 \tag{4.5}$$

$$r_2 = F_2^H y_2 = F_2^H H_2 (W_1 s_1 + W_2 s_2) + F_2^H z_2. \tag{4.6}$$

Block diagonalization was mentioned as a well-established scheme for MU-MIMO. For block diagonalization, the precoders should satisfy $H_2 W_1 \sim 0$ and $H_1 W_2 \sim 0$. However it is UE 1 that computes (and feeds back) W_1. It has no knowledge of the co-scheduled UE downlink channel H_2 in order to choose W_1 as per the criterion $H_2 W_1 \sim 0$. The best the UEs can do is choose the precoders as per the Release 8 SU-MIMO criterion. The eNodeB then picks up two UEs for MU-MIMO co-scheduling, whose reported PMIs correspond to orthogonal (or nearly orthogonal) precoders. This would ensure that the block diagonalization conditions are also satisfied.

For CQI reporting, the received SINR depends on the interference from the co-scheduled UE, i.e. the aforementioned $H_2 W_1$ and $H_1 W_2$ terms for UE 2 and UE 1 respectively. Since during CQI computation, UE 1 does not know the value of W_2 that the eNodeB would use during subsequent MU-MIMO transmission, it cannot accurately factor in the interference. Various proposals were submitted during 3GPP Release 10 standardization [6, 7, 8, 9] to address this issue. These proposals could be broadly categorized into two classes as follows.

1. The UE tries to estimate the CQI for MU-MIMO transmission under certain assumptions about the co-scheduled UE's precoder and reports an additional MU-CQI. For purposes of MU-CQI computation, the UE could, for example, assume that the eNodeB only schedules rank 1 transmissions (to itself and other co-scheduled UEs) and the other co-scheduled UEs have orthogonal precoders to the optimal precoder that it has computed.
2. The UE reports SU-MIMO CQI as before. If the eNodeB decides to pair this UE with another UE for MU-MIMO transmission, it recalculates the CQI that the UE can support based on the reported CQIs and PMIs of the two UEs.

Eventually MU-CQI reporting was not standardized in LTE as the associated gains were deemed insignificant. The UE still reports Release 8 PMI, CQI, and RI to keep the feedback framework common from Release 8 to Release 10. The eNodeB computes the MU-CQI from the UE feedback and UE pairing for MU-MIMO and reports the updated MCS to the UEs.[1]

4.4.2   Feedback for MIMO OFDM Channels

Guidelines for deriving CQI, PMI, and RI in narrowband MIMO channels were provided in the previous section. In practice, LTE is a wideband channel and a single CQI and PMI/RI will

---

[1] Change within 3GPP standardization is gradual and incremental. When progressing from one standard release to the next, *not* changing a lot of features simultaneously is a priority, hence the non-introduction of MU-CQI reporting by UEs. Another example is base station cooperation (CoMP) which was extensively discussed during the LTE Release 10 standardization phase, but not adopted. The discussions were carried over to the Release 11 standardization phase and basic CoMP features were adopted in Release 11. MU-CQI reporting by UEs could yet be adopted in a future LTE release.

not capture all the frequency selective information. In LTE, a UE reports CSI to the granularity of a group of tones called sub-bands. The sub-band size (number of tones) is a function of the system bandwidth. For each sub-band, the UE computes a single precoder and CQI value. The precoder could be chosen by maximizing the average channel norm (as defined in Equation (4.4)) over all the tones. The CQI is selected by assuming a single modulation and coding value for block(s) of bits that are mapped to the PDSCH REs of the sub-band, such that the CQI value corresponds to the highest MCS index that achieves the 10% BLER criterion. In addition to sub-band reporting, the UE can also report wideband CSI where a single CQI and PMI is reported assuming that the transmission to the UE spans the entire bandwidth. Note that during subsequent PDSCH transmission, if the UE is allocated sub-bands, the eNodeB calculates the value of a single MCS over the total frequency band. Frequency-selective MCS selection is not performed as the gains over single MCS is minimal and comes at a cost of high control overhead.

Some questions naturally arise at this point:

1. how frequently should the UE report CSI information to the eNodeB?
2. does the UE report all kinds of CSI (wideband and sub-band) at each feedback reporting instance?

The answers to these questions can be found in the LTE design principles of striking a reasonable trade-off between performance improvement and reducing design/feedback complexity and overheads. Accordingly, the following two types of feedback are supported by LTE.

- *Periodic*: This is transmitted on PUCCH. The reporting period is RRC configured. Since this is periodic and a PUCCH resource is made of only a few RBs, the UE typically transmits coarser feedback information. Periodic feedback enables the eNodeB to decide if it needs more detailed feedback from the UE. For example, the coarser information may make eNodeB *interested* in scheduling the UE. In order to do so, the eNodeB may ask for more detailed information via aperiodic feedback.
- *Aperiodic*: This is transmitted on PUSCH and typically has more detailed feedback information than periodic. The UE transmits aperiodic feedback when it receives a eNodeB trigger carried in DCI formats 0 or 4. Since a PUSCH resource to a UE can span much more RBs than PUCCH, it can hold more finer and detailed feedback information.

The terms *coarse* and *fine* feedback require to be defined. First, note that sub-band CSI feedback is finer than wideband feedback. Even for sub-band feedback, there are two kinds: UE selected and network (RRC) configured. Network-configured sub-band feedback is supported only for aperiodic feedback and is shown in Table 4.6. The UE reports a wideband CQI value and a differential CQI value assuming transmission over each sub-band.

UE-selected sub-band configuration differs from aperiodic to periodic feedback. For aperiodic feedback, the UE selects a set of M preferred sub-bands of size k. The UE reports a wideband CQI and one sub-band CQI value for each of these M sub-bands. The UE also reports the position of these M selected sub-bands. The configurations of k and M depend on system bandwidth, as shown in Table 4.7.

For periodic sub-band feedback the basic idea is still selective feedback of certain UE selected sub-bands; the implementation is slightly different, however. The total number of sub-bands N is divided into J groups called *bandwidth parts*. The UE selects a sub-band from

Table 4.6 Network-configured aperiodic sub-band feedback

System bandwidth (RBs)	Sub-band size (k RBs)
6–7	(Wideband CQI only)
8–10	4
11–26	4
27–63	6
64–110	8

(Source: 3GPP 2012 [9]. Reproduced with permission of 3GPP.)

Table 4.7 UE-selected aperiodic sub-band feedback

System bandwidth (RBs)	Sub-band size (k RBs)	No. of preferred sub-bands M
6–7	(Wideband CQI only)	(Wideband CQI only)
8–10	2	1
11–26	2	3
27–63	3	5
64–110	4	6

(Source: 3GPP 2012 [9]. Reproduced with permission of 3GPP.)

Table 4.8 UE-selected periodic sub-band feedback

System bandwidth (RBs)	Sub-band size (k RBs)	No. of bandwidth parts J
6–7	(Wideband CQI only)	1
8–10	4	1
11–26	4	2
27–63	6	3
64–110	8	4

(Source: 3GPP 2012 [9]. Reproduced with permission of 3GPP.)

each bandwidth part and reports its CQI along with its index. The values of k and J depend on the system bandwidth and are listed in Table 4.8.

4.4.2.1 Aperiodic and Periodic Feedback Payloads

To summarize the discussions of the previous section, there are different levels of feedback granularity associated with both periodic and aperiodic feedback. These are formally categorized under *CSI reporting modes*, listed in Table 4.9. For each UE, the network RRC configures a reporting mode. Some of the features of these modes are as follows.

1. Network-configured sub-band feedback for aperiodic reporting reports CSI of more sub-bands than UE-selected sub-bands.

Table 4.9   PUSCH and PUCCH CSI reporting modes

	No PMI	Single PMI	Multiple PMI
Wideband CQI	1-0 (PUCCH)	1-1 (PUCCH)	1-2 (PUSCH)
UE-selected sub-band CQI	2-0 (Both)	2-1 (PUCCH)	2-2 (PUSCH)
RRC-configured sub-band CQI	3-0 (PUSCH)	3-1 (PUSCH)	3-2 (In future)

2. For the UE-selected sub-band selection modes, the UE reports more sub-bands for aperiodic transmission than periodic transmission (i.e. compare the values of M in Table 4.7 to the corresponding values of J in Table 4.8).
3. The modes that report multiple PMIs are available only for aperiodic transmission.

This demonstrates the fact that periodic feedback reporting in PUCCH transmits coarser information than aperiodic feedback in PUSCH. The PUSCH 3-2 mode that would have reported the most detailed CSI information has been discussed in detail during past 3GPP standardization meetings. It has not yet been included in the standard however, as a consensus could not be reached as to whether the extra gains of PUSCH 3-2 are significant enough to justify the increased feedback complexity.

For aperiodic feedback, the UE is RRC-configured to one of the PUSCH CSI reporting modes listed in Table 4.9. It feeds back CQI, PMI, and the corresponding RI on the same PUSCH resource that has been assigned to it.

Even the reduced feedback payload of the various PUCCH reporting modes is not low enough, however. This is addressed by designing PUCCH reporting so that not all the information (CQI, PMI, and RI) need be sent simultaneously, i.e. over the same PUCCH resource. Each can be sent with different periods and offsets in different PUCCH resources. For example, in PUCCH mode 2-1, the UE should report a wideband CQI, sub-band CQIs, and a wideband PMI. The UE splits this information into two kinds of reports covering multiple reporting instances. One kind of report is for sub-band CQI and the other is for wideband CQI and PMI information. The CSIs corresponding to the J bandwidth parts for sub-band feedback could be reported in J different time instants between two consecutive wideband CQI/PMI reporting instances.

This is formally done by defining the following four *CSI reporting types* for each periodic reporting mode:

- *Type 1*: CQI feedback for the UE-selected sub-bands;
- *Type 2*: wideband CQI and PMI feedback;
- *Type 3*: RI feedback; and
- *Type 4*: wideband CQI.

As with CSI reporting modes, the CSI reporting types are also RRC configured for each UE. For each combination of reporting modes and reporting types, the content, payload (in terms of bits), periods, and offsets of the information to be reported are defined in [5]. The above example corresponded to a case when the UE had been configured in PUCCH mode 2-1 and reporting types 1 and 2. Note that all possible combinations of reporting modes and types are not possible.

Table 4.10 Transmission mode for PDSCH

Transmission mode #	Description
1	Transmission from a single antenna port
2	Transmit diversity
3	Open-loop spatial multiplexing
4	Closed-loop spatial multiplexing
5	MU-MIMO
6	Closed-loop rank 1 precoding
7	Transmission using DMRS with a single spatial layer
8	Transmission using DMRS with up to 2 spatial layers
9	Transmission using DMRS with up to 8 spatial layers
10	Coordinated multipoint (CoMP)

Finally, recall the PUCCH formats that were introduced in Chapter 2. Recall that PUCCH formats 2a and 2b were defined for carrying CSI. The PUCCH formats define the payload size and the modulation and coding used for transmission. The content of the payload in the PUCCH formats are determined by the PUCCH feedback modes discussed above.

4.4.2.2 Transmission Mode

In LTE the term transmission mode (TM) is used to describe the way in which data are transmitted from multiple antennae and how data are subsequently demodulated at the UE side. Different transmission modes may differ in the signal processing used at the eNodeB (spatial multiplexing versus diversity) or in the demodulation used at the UE (use of CRS or UE-specific DMRS). The TMs for PDSCH transmission are listed in Table 4.10. TM 1 uses only one antenna port. TM 2 is the default MIMO mode, where Space Frequency Block Code (SFBC) is used to achieve transmit diversity. In TM 3, the same OFDM signaling is mapped to each antenna port with a specific delay. TM 4 supports spatial multiplexing with up to four antenna ports. In this mode, the UE measures the CSI based on the CRS and feeds back the Rank Indicator (RI), Precoding Matrix Indicator (PMI), and Channel Quality Indicator (CQI), which will be used by the eNodeB to choose appropriate transmission parameters. The Multi-User MIMO (MU-MIMO) is supported in TM 5. TM 6 is a special type of TM 4, where only one layer is allowed for data transmission. Single-layer beamforming and dual-layer beamforming are supported by TM 7 and TM 8, respectively. In these two modes, the CRS and DMRS are used for CSI measurement and demodulation, respectively. The beamforming technique is further extended in TM 9 to support up to 8-layer transmission where the CSI-RS and DMRS are used for CSI measurement and demodulation, respectively. TM 10 supports Coordinated Multi-Point (CoMP) transmision/reception, which will be explained later in Chapters 8 and 9. The eNodeB configures different TMs from Table 4.10 for different UEs via RRC signaling.

4.4.3 New Features in LTE-Advanced

LTE Release 10 introduced eight antenna transmissions and the subsequent need for new codebooks for eight antenna ports. A dual codebook structure was introduced with two codebooks: $\mathbf{W}_1$ which is aligned to the long-term wideband correlation of the channel and $\mathbf{W}_2$ which is aligned to the frequency-selective sub-band properties of the channel. Since $\mathbf{W}_1$ carries wideband information that changes infrequently, it can be calculated and reported over longer time intervals than the sub-band precoder $\mathbf{W}_2$. This reduces feedback overheads.

After much discussion during the standardization phase, the final effective codeword was chosen to be of the form $\mathbf{W} = \mathbf{W}_1 \mathbf{W}_2$. For rank r transmission, matrix $\mathbf{W}_1$ is a *tall* precoder of dimension $N_T \times n$ with $n < N_T$ and $\mathbf{W}_2$ is of rank $n \times r$. The dual codebook structure of Release 10 has been designed to be optimized for cross-polarized deployments, the dominant deployment scenario of the future.

We now provide insights into designing the dual codebooks. Consider a cross-polarized channel matrix $\mathbf{H}$ which can be expressed as a block matrix $\mathbf{H} = [\mathbf{H}_{\mathrm{cpol}}, \mathbf{H}_{\mathrm{xpol}}]$ where each sub-block is of dimension $N_R \times N_T/2$ and corresponds to a particular polarization. The precoder $\mathbf{W}_1$ is chosen to be a block diagonal of the form $\mathbf{W}_1 = \mathrm{diag}\{\widetilde{\mathbf{W}}_1, \widetilde{\mathbf{W}}_1\}$ such that $\widetilde{\mathbf{W}}_1$ is of dimension $N_T/2 \times n/2$. For this structure we have

$$\mathbf{HW} = [\mathbf{H}_{\mathrm{cpol}}, \mathbf{H}_{\mathrm{xpol}}]\mathbf{W}_1 \mathbf{W}_2 \tag{4.7}$$

$$= [\mathbf{H}_{\mathrm{cpol}}, \mathbf{H}_{\mathrm{xpol}}] \begin{bmatrix} \widetilde{\mathbf{W}}_1 & 0 \\ 0 & \widetilde{\mathbf{W}}_1 \end{bmatrix} \mathbf{W}_2 \tag{4.8}$$

$$= [\mathbf{H}_{\mathrm{cpol}}\widetilde{\mathbf{W}}_1, \mathbf{H}_{\mathrm{xpol}}\widetilde{\mathbf{W}}_1]\mathbf{W}_2 = \mathbf{H}_{\mathrm{eff}}\mathbf{W}_2. \tag{4.9}$$

This gives an insight into precoder selection from a codebook. First a precoder from $\widetilde{\mathbf{W}}_1$ matching the ULA correlations can be seen in $\mathbf{H}_{\mathrm{cpol}}$. Next $\mathbf{W}_2$ can be chosen to select columns from the effective channel $\mathbf{H}_{\mathrm{eff}}$ and perform phase adjustments. The following example illustrates the process of codebook design.

Example 4.4.1 *Consider a special case of $n = 2$ and $r = 1$. $\widetilde{\mathbf{W}}_1$ is a $N_T \times 1$ vector, denoted* w. $\mathbf{W}_2$ *is a 2×1 vector, denoted $[1, \alpha]^T$. The overall precoder is therefore of the form:*

$$\mathbf{W} = \begin{bmatrix} \mathbf{w} & 0 \\ 0 & \mathbf{w} \end{bmatrix} \begin{bmatrix} 1 \\ \alpha \end{bmatrix}. \tag{4.10}$$

From the principles of codebook design for Release 8, the vector w *can be chosen as $a_{N_T/2}(\theta)^T$ and then α can be chosen to phase adjust between the two polarizations.*

Although similar to Release 8 design, this has two potential benefits as follows.

1. The phase adjustment α could be performed separately for each sub-band. This flexibility was not present in Release 8, where a single precoder had to be chosen for all allocated sub-bands.
2. For a total PMI feedback budget of K bits, the bits allocated for $\mathbf{W}_1$ and $\mathbf{W}_2$ can be adaptively chosen to suit the specific transmission environment.

In the actual Release 10 precoder, $\widetilde{\mathbf{W}}_1$ has multiple columns and the $\mathbf{W}_2$ operation selects certain column vectors from $\widetilde{\mathbf{W}}_1$ and then performs phase adjustment [1]. Although Release 10 is optimized for a cross-polarized set-up, with proper selection of $\mathbf{W}_2$ it also works well for co-polarized settings.

Feedback reporting for Release 10 extends the Release 8/9 reporting principles to incorporate the dual codebook feedback. Newer feedback report formats are configured that define new UE behaviors in which the UE reports a subset of W_1/W_2 or both wideband/subband CQI and RI. Thus W_1 and W_2 may be reported in the same or different subframes. The fundamental principle is still to design multiple feedback modes such that in each mode the UE reports partial information to save on feedback overhead. It is the task of the eNodeB to configure the UE in these various feedback modes to extract information.

Moving ahead, LTE Release 11 has devoted much attention to heterogeneous networks. Coordination between macro- and piconodes/RRHs to serve a UE in the downlink is possible (CoMP) as a UE can receive PDSCH from multiple points (eNodeBs) dynamically. Feedback signaling has been accordingly enhanced as a UE now needs to feed back CSI for all eNodeBs (macro and pico/RRH) that can potentially serve it. This is discussed in more detail in Chapter 9.

4.5   Design Paradigm of LTE Signaling

In the last two chapters we have covered various aspects of LTE signaling, including DCI formats, LTE transmission modes, and PUSCH and PUCCH CSI reporting modes. In this section, we shall briefly demonstrate the link between these aspects and how they evolve during the process of standardization.

When a new feature is considered for introduction in LTE, the first question is how it would affect network and UE behavior. Examples of UE behavior could be a UE supporting MU-MIMO or CoMP reception or the eNodeB and UE deploying eight antennae. Let us start with the eight antenna deployment as an example, introduced during Release 10 standardization phase. To enable this, the UE needs to feed back the dual codebook-based CSI. A UE should also be capable of demodulating PDSCH that can now be up to eight layers for SU-MIMO. It was also decided that MU-MIMO should be an important application area for eight antenna deployments due to the increase in spatial layers.

A new UE *behavior* needed to be defined. A behavior of a UE is defined by issues such as how and what it reports as CSI, how the eNodeB transmits PDSCH to it, and how it demodulates the PDSCH (CRS or DMRS), etc. As explained in Chapter 3, different UE behaviors are formally captured in the transmission modes. Thus for eight antenna communications, a new transmission mode was first defined and called TM 9. In TM 9, the PUSCH and PUCCH reporting defined in Release 8 were extended to cover the reporting of dual PMI. Since CRS is not defined for eight antenna ports, new CSI-RS patterns for eight antenna ports were defined and associated with TM 9. The UE may also now receive up to eight-layer PDSCH for SU-MIMO and four-layer PDSCH for MU-MIMO. The UE is transparent to the SU/MU-MIMO operation. To enable this, new control signaling was designed and thus a new DCI format 2C was developed for TM 9.

Once a UE/network behavior is identified, a new transmission mode is usually defined which may also need changes to channel measurement and feedback procedures and a new DCI format [10].

4.6   Scheduling and Resource Allocation

The eNodeB in a LTE network is the only entity between the users and the core network that is responsible for the provision of guaranteed quality of service (QoS) to all the users in its

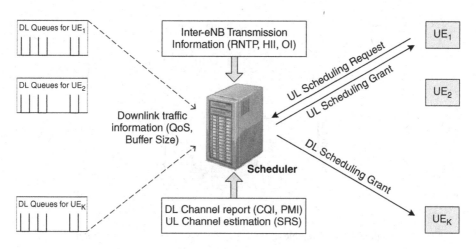

Figure 4.12 Radio resource scheduler in an LTE network

cell. The eNodeB must perform efficient radio resource allocation both in the downlink and the uplink channels that satisfies the majority of the users in terms of various QoS parameters such as delay, jitter, throughput, or outage probability. The eNodeB ensures traffic differentiation by mapping the downlink traffic for a given UE into different queues, each with a QoS constraint corresponding to the requirements of the application. Similarly, for the uplink transmission each UE maintains multiple queues with differing QoS constraints. The objective of the resource scheduler at the eNodeB is to allocate radio resources both in the uplink and the downlink directions that meet the QoS requirements for the majority of the users. Figure 4.12 shows an LTE scheduler serving multiple UEs both in the uplink and the downlink directions.

The radio links between an eNodeB and the UEs are subjected to fluctuating link error rates and capacities due to time-varying fading, noise, and co-channel interference arising from the neighboring cells. The scheduling algorithms at the eNodeB are required to ensure the guaranteed QoS service parameters despite the unreliable nature of the wireless medium. An LTE resource scheduler allocates the bandwidth resources to a selected set of UEs in each subframe along with corresponding transmission powers in order to optimize the overall QoS of the served UEs. The scheduling decision is often based on and not limited to the underlying radio link quality of different UEs, queue buffer sizes, QoS constraints, status of the previous transmissions, and interference condition in the neighboring cells.

As shown in Figure 4.12, the resource scheduling process is tightly coupled with the channel measurements provided by the UEs. For the downlink channel all UEs periodically report channel quality indicator (CQI) as discussed in Section 4.4. Similarly, to support the uplink transmissions, the eNodeB periodically schedules the UEs to transmit sounding reference signals (SRS) which provide the eNodeB with an estimate of the uplink channel quality. In addition to the channel measurements, the scheduling decision is also based on the underlying link adaptation scheme and the retransmission protocol. For example, a successful transmission depends not only on the radio link quality but also on the payload size it carries, the modulation and coding schemes used, and the packet combining scheme used by the UE in case of retransmissions. In LTE, a Hybrid Automated Repeat Request (HARQ) based retransmission protocol is used which significantly improves the spectral efficiency by combining the

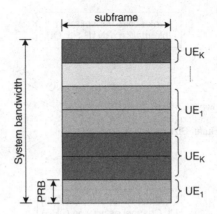

Figure 4.13   Example of frequency diversity in resource allocation

forward error-correcting codes present in each retransmission. The use of HARQ ensures that the probability of successfully decoding a packet increases with each subsequent retransmission. A smart scheduler therefore makes the scheduling decision not only based on prevalent channel condition but also takes the retransmission attempt into consideration.

Figure 4.13 depicts an example of a typical RB allocation in LTE. As not all UEs can be served at a given instant, the scheduler exploits multi-user diversity when selecting UEs for scheduling. In addition to multi-user diversity, the scheduler also exploits frequency diversity as different UEs undergo different fading in different RBs. The RBs allocated to a UE are therefore not contiguous but spread across the available system bandwidth. However, a constraint of LTE is that the same MCS value for a given codeword spanning multiple RBs is used, irrespective of the reported CQIs. This constraint was adopted as the RB-specific choice of MCS was deemed to offer negligible gains over single MCS choice while increasing the feedback overheads in the other direction. The use of codeword-specific MCS is usually beneficial and is allowed in LTE, however. It may be noted that the scheduler not only selects the UEs but also decides the location of allocated RBs, choice of MCS, and the transport block size. In addition, LTE Release 10 also supports scheduling of multiple UEs in the same time/frequency resources, utilizing advanced MIMO signal processing.

4.6.1   Scheduling Algorithms

The choice of scheduling algorithm is critical in provision of the QoS to all the users in the network. There exist several classes of scheduling algorithms that can be considered for a network based on the availability of traffic information, QoS requirements, and the measure of radio link qualities for all the users. One of the most well-known scheduling algorithm is the 'round robin' (RR) algorithm that schedules users in a given order without requiring any additional user-specific information. An RR scheduler guarantees the fairness to all users by providing equal radio resources and transmission opportunities to all the users. However, RR is suitable only for a transmission medium where users have a relatively similar radio link quality at any given time, such as the ethernet. RR is rarely used in a cellular wireless network where variable link error rates are encountered for users based on their surroundings, proximity to the transmitter, and mobility.

The maximum rate (MR) algorithm overcomes the disadvantages of an RR scheduler by utilizing channel state information. It allocates radio resources to a subset of users with best channel state in given time/frequency resources. For a set of achievable rates, $r_i(t)$ of UE i, in a given time/frequency resource at time t, MR schedules a UE that offers the best data rate among all other UEs, that is,

$$k = \arg \max r_i(t). \tag{4.11}$$

Since MR is channel-aware scheduler, it can provide significantly better throughput performance when compared to the RR scheduler. In fact MR is the optimal scheduler for system throughput maximization. However, an MR scheduler is biased against users with poor average radio link quality and thus cannot guarantee fairness to all users, as depicted in Figure 4.14. An MR scheduler is therefore not suitable for a cellular network where average link quality varies markedly for users near the base station and those at the cell edges.

A proportional fair (PF) scheduler strikes a reasonable trade-off between system throughput optimization and fair allocation of resources to all users. Instead of making scheduling decisions based on the absolute channel quality of users, a PF algorithm considers the weighted channel condition of all the users. The following criteria describes the user selection for a PF scheduler [11]:

$$k = \arg \max \frac{r_i(t)}{R_i(t)} \tag{4.12}$$

where $R_i(t)$ is the average data rate for the ith user defined over a time window (t_c) as

$$R_i(t) = \left(1 - \frac{1}{t_c}\right) R_i(t-1) + \frac{1}{t_c} r_i(t-1). \tag{4.13}$$

PF scheduling ensures that users with poor radio link quality are not ignored in the scheduling process. Figure 4.14 provides an example of user selections for MR and PF algorithms. It can be observed that, unlike MR, PF does well on fairness criteria by allocating resources to UE_3 when its SINR is above the average SINR experienced by it. Note that the window

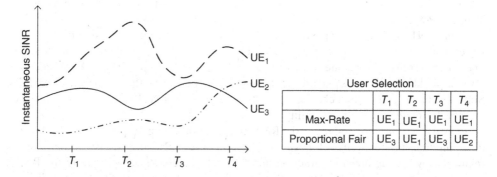

Figure 4.14 Scheduling choices based on different scheduling algorithms

size t_c determines the trade-off between the achievable system throughput and the fairness. For $t_c \gg 1$, the scheduling decision is dictated by the instantaneous channel condition. PF therefore converges to MR for large values of t_c. At the other extreme, PF converges to RR for smaller value of t_c.

One of the shortcomings of the PF scheduler is that it bases scheduling decisions solely on the channel quality while ignoring QoS constraints such as delay. Hence, the PF algorithm described in Equation (4.12) is not suitable for real-time multimedia traffic which has strict delay and jitter requirements. A more general PF selection metric can be defined as:

$$k = \arg \max \frac{r_i(t)}{R_i(t)} \alpha_i \tag{4.14}$$

where α_i represents the effective QoS metric for the ith user. One such variant of a PF scheduler is maximum-largest weighted-delay first (M-LWDF) algorithm which defines α_i [12] as:

$$\alpha_i = -\frac{\log (\delta_i)}{\tau_i} W_i(t) \tag{4.15}$$

where for the ith user $w_i(t)$ denotes the current waiting time for the next available packet since its arrival in the queue; τ_i is the delay threshold; and δ_i is the maximum probability that $w_i(t)$ exceeds τ_i. Thus, M-LWDF not only guarantees the fairness but also incorporates delay parameter to ensure guaranteed QoS for real-time traffic.

4.6.2   Inter-eNodeB Coordination for Resource Allocation in LTE

In a cellular network, the resource scheduling decision in one cell impacts its neighboring cells' performance due to the co-channel interference caused in both the downlink and the uplink channels. An ideal resource allocation scheduler should take the resulting interference into account. In LTE Release 8 there are mechanisms for the exchange of limited interference information among the neighboring eNodeBs via the X2 interface for frequency-domain inter-cell interference coordination (ICIC). The X2 interface serves as the communication link for signaling exchange among the neighboring eNodeBs. A good resource scheduler must make use of the interference-related information obtained over the X2 interface to improve overall system performance.

For the downlink transmissions, an eNodeB can utilize a bitmap called Relative Narrowband Transmit Power (RNTP) to inform the neighboring eNodeBs of its transmission power. Each bit in the RNTP bitmap corresponds to a RB and is set to 1 if the transmission power in that RB exceeds a predefined threshold normalized by the cell size. The normalized threshold provides a fair estimate of the interference at the cell edge for a macrocell as well as a picocell. The threshold value is not standardized and can be negotiated by the neighboring eNodeBs. Similarly, the behavior of an eNodeB upon the receipt of the RNTP message is not designed to allow for more scheduling flexibility.

Figure 4.15 depicts an example of the RNTP information where the pico eNodeB schedules downlink transmission to its cell-edge UE when the macro eNodeB's transmission power is relatively low during downlink transmission to a cell-center UE. In addition to the RNTP, Release 10 introduced additional mechanisms for time-domain interference coordination involving transmission of an Almost Blank Subframe (ABS) by the macro eNodeB to protect

(a)   (b)

Figure 4.15 Scheduling decisions based on interference coordination between a macrocell and a picocell

the downlink transmission to a pico UE on the cell edge. The details of ABS mechanism are explained in Chapter 5.

Similar to the co-channel interference caused by the downlink transmissions, uplink transmissions by cell-edge UEs can also cause considerable interference at the neighboring eNodeBs. This becomes a severe issue if the cell-edge UEs at the neighboring cells are co-scheduled with the same RB resources. To minimize such occurrences, LTE allows the neighboring eNodeBs to share interference levels corresponding to each RB via an Overload Indicator (OI) message. The OI bitmap can be shared with a periodicity of 20 ms or more. As in the case of RNTP, LTE does not specify the behavior of an eNodeB in response to the OI messages. For example, upon receiving OI messages indicating high interference in certain RBs, the aggressor eNodeB may choose to avoid scheduling cell-edge UEs in those RBs, instead allocating those resources to cell-center UEs. It may also choose to instruct UEs to reduce transmit power to minimize interference to the neighboring cells. In addition to OI messages, LTE also enables an eNodeB to advertise likely uplink interference resulting from scheduling of cell-edge UEs in future subframes. This is accomplished by sending out a High Interference Indicator (HII) over the X2 interface. Exchange of OI and HII messages is useful in enabling eNodeBs to make informed uplink scheduling decisions.

Interference coordination schemes based on the exchange of RNTP, OI, and HII messages ensure that the neighboring eNodeBs do not schedule their cell-edge UEs simultaneously over the same RBs, thus avoiding co-channel interference in the downlink and uplink channels. In addition to these interference coordination mechanisms, LTE Releases 10 and 11 introduced several advanced features such as enhanced ICIC (eICIC) and Coordinated MultiPoint (CoMP) to enable multiple eNodeBs to perform collaborative resource scheduling to further improve the throughputs of their cell-edge UEs.

An example of a CoMP scheme is Dynamic Point Selection (DPS), which allows for greater scheduling flexibility among collaborating eNodeBs. In DPS a cell-edge UE of one cell may be served by a neighboring eNodeB depending upon better radio link quality or load balancing. Similarly, another CoMP scheme called Joint Transmission (JT) can provide improved throughput via joint transmission/reception from/at multiple eNodeBs.

Figure 4.16 depicts an example where a macro eNodeB chooses a CoMP UE for downlink transmission instead of its own UE. The performance improvement offered by CoMP schemes comes at the expense of increased signaling among different schedulers as well as the exchange

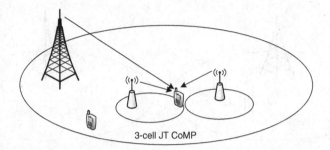

3-cell JT CoMP

Figure 4.16   An example of JT CoMP for the downlink channel between the macrocell and picocells

of user data at multiple eNodeBs. More details on CoMP features are provided in Chapters 8 and 9.

References

[1] 3GPP (2012), Evolved Universal Terrestrial Radio Access (E-UTRA), Physical channels and modulation 3GPP TS 36.211 v11.0.0. Third Generation Partnership Project, Technical Report.

[2] DoCoMo (2012) R1-121935, Interference Measurement Mechanism for Rel-11. CoCoMo, Technical Report, May 2012.

[3] Love, D., Heath, R., Lau, V.K.N., Gesbert, D., Rao, B.D., and Andrews, M. (2008) An overview of limited feedback in wireless communication systems. *IEEE Journal on Selected Areas in Communications*, **26**(8), 1341–1365.

[4] Love, D., Heath, R., and Strohmer, T. (2003) Grassmannian beamforming for multiple-input multiple-output wireless systems. *IEEE Transactions on Information Theory*, **49**(10), 2735–2747.

[5] 3GPP (2012) Evolved Universal Terrestrial Radio Access (E-UTRA), Physical layer procedures 3GPP TS 36.213 v11.0.0. Third Generation Partnership Project, Technical Report, September 2012.

[6] Huawei, HiSilicon (2010) R1-105133, CQI Enhancement for Rel-10 MU MIMO. Huawei, Technical Report, October 2010.

[7] Marvel (2010) R1-105752, CQI Enhancement for 4Tx. Marvel, Technical Report, October 2010.

[8] Texas Instruments (2010) R1-105284, Rel.10 CSI Enhancement for MU-MIMO. Texas Instruments, Technical Report, October 2010.

[9] Nokia, Nokia Siemens Networks (2010) R1-105532, On the MU-MIMO Feedback Enhancements for LTE-Advanced. Nokia, Technical Report, October 2010.

[10] 3GPP (2012) Evolved Universal Terrestrial Radio Access (E-UTRA) and Evolved Universal Terrestrial Radio Access (E-UTRAN), overall description 3GPP TS 36.300 v11.2.0. Third Generation Partnership Project, Technical Report, June 2012.

[11] Almatarneh, R.K., Ahmed, M.H., and Dobre, O.A. (2010) Performance analysis of proportional fair sceduling in OFDMA wireless systems. In *Proceedings of IEEE Vehicular Technology Conference (VTC)*, pp. 1–5, September 2010.

[12] Andrews, M., Kumaran, K., Ramanan, K., Stolyar, A., Whiting, P., and Vijayakumar, R. (2001) Providing quality of service over a shared wireless link. *IEEE Communications Magazine*, **39**(2), 150–154.

Part Two

Inter-Cell Interference Coordination

5

Release 10 Enhanced ICIC

5.1 Introduction

In a LTE heterogeneous network, small cells such as pico- and femtocells are deployed within the coverage area of macrocells to increase network capacity and handle non-uniform UE traffic distribution. The deployment of small cells potentially enhances the throughput performance by providing extra cell-splitting gain, especially for the UEs on macrocell edges and in hotspots. On the other hand, it may lead to increased interference in the network and thus reduce UE throughput. The interference is more significant when the small cells are deployed on the same carrier as the macrocells. It is therefore important to coordinate the interference between the macro- and small cells in order to maximize the benefit of deploying small cells.

In this chapter, we explain the details of the enhanced Inter-Cell Interference Coordination (eICIC) techniques that are standardized in LTE-Advanced Release 10 for small cell deployments. In Section 5.2, we introduce two typical small cell deployment scenarios: the macro–pico deployment scenario and the macro–femto deployment scenario. We then discuss the time domain and power control-based eICIC techniques in Section 5.3 and Section 5.4, respectively. Finally we explain the eICIC techniques based on Carrier Aggregation (CA) in Section 5.5.

5.2 Typical Deployment Scenarios

One of the most important features in LTE Release 10 is the support of heterogeneous network deployments, in which macrocells provide the basic coverage and small cells are deployed within the coverage area of macrocells to further enhance UE throughput. The following two typical small cell deployment scenarios are considered in Release 10: the macro–pico deployment and the macro–femto deployment. The basic difference between the two scenarios is that femtocells cannot be accessed by all UEs in the network and femto UEs form a closed subscriber group (CSG), whereas picocells are open to all UEs. These two deployment scenarios have different interference situations and backhaul supports, which require different techniques for interference management.

Heterogeneous Networks in LTE-Advanced, First Edition. Joydeep Acharya, Long Gao and Sudhanshu Gaur.
© 2014 John Wiley & Sons, Ltd. Published 2014 by John Wiley & Sons, Ltd.
Companion Wesite: www.ltehetnet.com.

5.2.1 Macro–Pico Deployment Scenario

In the macro–pico deployment scenario, picocells are deployed by the network operator in a planned way. For example, they can be placed on the edge of a macrocell or in a hotspot to enhance UE throughput. A Pico eNodeB (PeNodeB) that forms a picocell has the same protocol stack and functionalities as a Macro eNodeB (MeNodeB), but transmits with lower power.

The network architecture of the macro–pico deployment scenario is depicted in Figure 5.1. Both eNodeBs are connected to a Mobile Management Entity (MME) and a Serving Gateway (S-GW) via the S1 interface. They are also connected to their neighboring MeNodeBs and PeNodeBs via X2-based backhaul. Both S1-based and X2-based UE handovers are therefore supported in this deployment scenario. As will be shown later in Section 5.3, the X2 interface also facilitates the interference coordination between a macrocell and a picocell.

In the macro–pico deployment scenario, a UE measures the downlink Reference Signal Received Power (RSRP) of its neighboring cells and chooses the one with the highest RSRP level as its serving cell. As a result, a picocell will serve a small number of UEs due to its low transmit power. In order to offload more data traffic to picocells, a logical bias can be added to the RSRP of the picocell before comparing it with the RSRP of another cell such as the macro. This biasing technique, called the Cell Range Expansion (CRE) bias, has been proposed to expand the serving area of a picocell without increasing its transmit power [1, 2, 3]. An example is shown in Figure 5.2 where PeNodeB 1 with positive CRE bias has a larger serving area than PeNodeB 2 which does not apply a CRE bias when the two PeNodeBs have the same transmit power.

How does CRE biasing work? In order to answer this question, we first briefly review the UE handover procedure. Figure 5.3 illustrates the X2-based handover procedure between a macro- and a picocell. As shown in the figure, a RRC_CONNECTED UE is initially connected to a MeNodeB. The UE measures the RSRP from its neighboring eNodeBs and reports these values to its associated MeNodeB periodically. The MeNodeB decides whether to hand

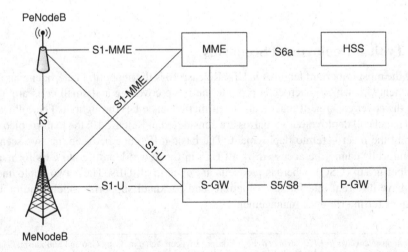

Figure 5.1   The architecture of the macro–pico deployment

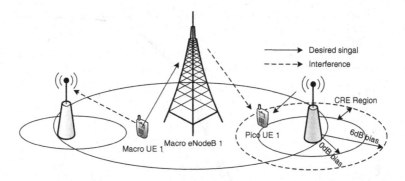

Figure 5.2 The interference situations in the macro–pico deployment scenario

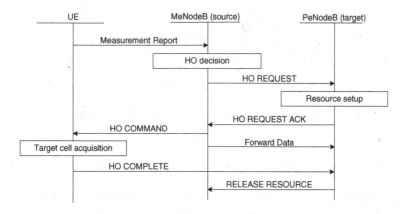

Figure 5.3 X2-based handover procedure

over the UE to other eNodeBs based on the received RSRP measurement report. If the RSRP of a neighboring PeNodeB becomes larger than that of the MeNodeB, the MeNodeB will initiate the handover process of the UE by sending a HANDOVER (HO) REQUEST message to the PeNodeB via the X2 interface. After receiving the HO REQUEST message, the PeNodeB prepares radio resources for the UE and replies back to the MeNodeB with a HO REQUEST ACK message. The MeNodeB then sends a HO COMMAND message to the UE in order to initiate the acquisition process of the picocell. Meanwhile, the MeNodeB starts data forwarding to avoid any data loss. After the UE completes the cell acquisition process, it informs the PeNodeB by sending a HO COMPLETE message. After receiving the HO COMPLETE message, the PeNodeB sends a RESOURCE RELEASE message to the MeNodeB via the X2 interface, which completes the whole handover process.

The CRE bias technique is implemented by the MeNodeB as part of the handover decision for a RRC_CONNECTED UE. The basic idea is to add a CRE bias to the RSRP of a picocell and compare the sum with the macro RSRP to decide if handover is to be performed. This way the MeNodeB can hand over a UE to a picocell with a relatively smaller RSRP value. This will consequently increase the coverage of the picocell. An example is shown in Figure 5.2 where

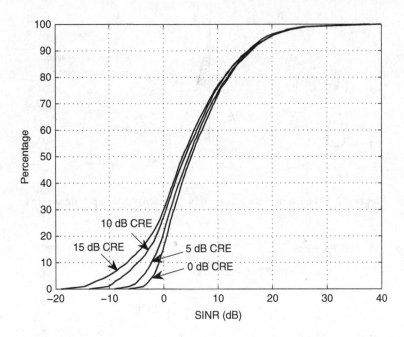

Figure 5.4   The CDF of SINR for pico UEs with different CRE bias values

the Pico UE 1 is served by PeNodeB 1 for 6 dB CRE bias but, in the absence of CRE bias, is served by MeNodeB 1.

The incremental increase in coverage of a picocell due to CRE is called the CRE region. A larger CRE bias will increase the CRE region and thus offload more UEs to the picocells. Note that CRE can only be applied to the UEs in the RRC_CONNECTED mode. The UEs in the RRC_IDLE mode will perform normal cell selection/reselection based on the unbiased RSRP. As a result, the RRC_IDLE UEs in the CRE region of a picocell are served by the cell with the highest RSRP, which is not that picocell.

In the macro–pico deployment scenario, interference management is a key issue especially when CRE is configured. The problem is most severe in downlink for the RRC_CONNECTED UEs in the CRE region of a picocell. In the example depicted by Figure 5.2, pico UE 1 suffers strong interference from MeNodeB 1, which can be up to 6 dB stronger than the desired signal from PeNodeB 1. In Figure 5.4, we show the Cumulative Distribution Function (CDF) for the received Signal-to-Interference-plus-Noise Ratio (SINR) of pico UEs with different CRE bias values in a typical macro–pico deployment scenario. In the evaluation, 57 macrocells are deployed uniformly following a hexagonal layout in the two-dimensional plane and 4 picocells are deployed randomly in each macrocell (refer to Section 5.3.5 for the detailed simulation assumptions). We see that the minimum value of UE SINR values decrease as the CRE bias increases. The UEs corresponding to these SINR values are located in the CRE region and face strong interference. As will be shown in Section 5.3, time domain eICIC techniques can be utilized to handle this interference situation.

In uplink transmit, power control is applied so that the received power at the desired cell meets the SINR threshold. Since UEs associated with the macrocell are located far away from it, they transmit with relatively higher power than the pico UEs which are located closer to the picocell. The interference that the macro UEs cause at the pico eNodeBs can therefore be severe. An example is depicted in Figure 5.2 where macro UE 1 causes interference at PeNodeB 2 when it transmits to MeNodeB 1. The uplink power control techniques in LTE Release 8/9 are sufficient to handle this interference situation [4].

5.2.2   Macro–Femto Deployment Scenario

In the macro–femto deployment scenario, the femtocells are usually deployed by the customer in an unplanned way. Femtocells can be placed inside an office building or a residential area to enhance UE throughput. A Home eNodeB (HeNodeB) that forms a femtocell is similar to a PeNodeB in the sense that both nodes have the full protocol stack and functionalities of a MeNodeB but transmit with lower power. The major differences between a HeNodeB and a PeNodeB are listed as follows.

- A HeNodeB is accessible only to a particular group of UEs in the network, i.e. the femtocells are Closed Subscriber Group (CSG) cells. This affects the network architecture and requires extra functionalities of the HeNodeB such as CSG provisioning and broadcasting. Furthermore, the mobility management of HeNodeBs is more complicated than that of picocells. The deployment of CSG cells also creates a unique interference situation in downlink.
- No direct backhaul exists between a HeNodeB and a MeNodeB; an X2-based handover between a MeNodeB and a HeNodeB is therefore not supported. Moreover, the interference coordination between a macrocell and a HeNodeB is more difficult due to the lack of direct backhaul.

The network architecture of the macro–femto deployment scenario is shown in Figure 5.5. A new network element, the CSG server, is adopted for the purpose of CSG provisioning which includes the management of the CSG data in the network as well as at the UE. The CSG server hosts functions for adding or deleting a subscriber from a CSG and for viewing the list of subscribers in a CSG. The CSG subscription data is stored in the Home Subscriber Server (HSS) and then transferred to the MME when a UE registers with the network.

The UE stores two kinds of CSG lists, called the *allowed* CSG list and the *operator* CSG list, in its SIM card. Both these lists contain entries of cells with their CSG Identity (ID) and associated Public Land Mobile Network (PLMN) ID. The difference between the two is that the allowed CSG list is managed by the UE and/or the network, while the operator CSG list is controlled only by the network. The CSG server updates these lists via the Open Mobile Alliance Device Management (OMADM) or Over The Air (OTA) procedures. The combination of the two CSG lists is called the CSG white-list. The UE considers the CSG IDs stored in the CSG white-list valid only within the scope of its registered PLMN.

In the macro–femto deployment scenario, a HeNodeB can be directly connected to a MME/S-GW via the S1 interface as for a PeNodeB in the macro–pico deployment scenario. It can also be connected to the MME/S-GW via a HeNodeB gateway (which is not shown in Figure 5.5). The HeNodeB gateway will reduce the signaling overhead that would otherwise have been propagated to the MME in the dense deployment of femtocells. A HeNodeB does not support the X2 interface in Release 8/9, which implies that only the S1-based handover is

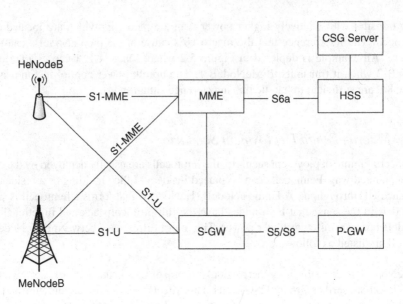

Figure 5.5   The architecture of the macro–femto deployment

supported from/to a femtocell. In Release 10, A HeNodeB can support the X2 interface with its neighboring HeNodeBs. However, the X2-based handover is supported only in the case where no CSG access control is necessary, i.e. the handover occurs between two femtocells with the same CSG ID.

The S1-based handover procedure in the macro–femto deployment scenario is shown in Figure 5.6. Initially the UE is served by a MeNodeB. The UE measures the RSRP from its neighboring macro-/femtocells and reports those values to the MeNodeB periodically. At this stage, the HeNodeB can tell whether a cell in the received measurement report is a CSG femto-cell or not by checking its Physical Cell Identity (PCI) included in the report.[1] If the RSRP of a HeNodeB becomes stronger than that of the MeNodeB, the MeNodeB will request the UE to report the CSG-related information of the HeNodeB for preliminary access control. The information includes the CSG ID of the target cell and the CSG member status which indicates whether the UE is a member of the CSG or not. The UE then performs cell acquisition, reads the CSG ID from the SIB1, and reports the requested information to the MeNodeB. Afterwards, the HeNodeB performs preliminary access control by verifying the CSG membership of the UE. If the verification passes, the MeNodeB includes the CSG ID in the HANDOVER (HO) REQUEST message to the MME.

In principle, the MME is responsible for CSG-related access control since the CSG sub-scription information of the UE, including the CSG white-list, is available in the MME. After receiving the HO REQUEST message, the MME will verify whether the UE is eligible to access the target HeNodeB and sends a HO REQUEST message to the target HeNodeB. After receiving the HO REQUEST message, the target HeNodeB will again verify whether

---

[1] The whole PCI range (from 0 to 503) is divided into a CSG range and a non-CSG range. Any CSG cell should be configured to use one of the PCIs in the CSG range.

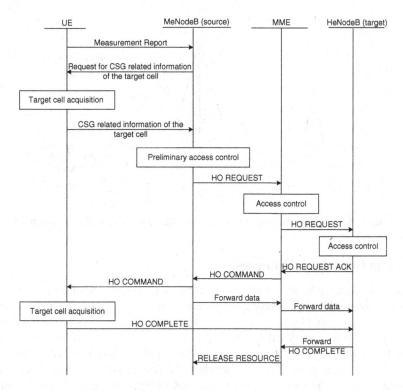

Figure 5.6 S1-based handover procedure

the CSG ID included in the HO REQUEST message is the same as that broadcast by itself. This verification is to prevent access attempts from a UE providing a fake CSG ID that can pass the MME CSG membership check. If the validation is successful, the target HeNodeB will prepare radio resource for handover and complete the whole process.

The interference issue is more severe in the macro–femto deployment scenario compared to the macro–pico deployment scenario due to the deployment of the CSG femtocells. Interference is more manageable in the macro–pico deployment scenario, since the UEs can be handed over freely between macro- and picocells. For example, if a macro UE is located on the edge of a macrocell and suffers strong interference from a nearby picocell, it can be handed over to the picocell. Such a handover may not be possible in the macro–femto deployment scenario, however. As shown in Figure 5.7, macro UE 1 will suffer strong interference from HeNodeB 1. If it does not belong to the CSG list, it cannot be handed over to HeNodeB 1. This can cause a *dead zone* around each CSG HeNodeB, within which macro UEs are not able to receive data packets from their serving macrocells. Such dead zones are larger if femtocells are deployed on the edge of a macrocell since the received signal from the macrocell is weaker. The limited interference coordination between a femtocell and a macrocell due to the lack of direct backhaul makes the problem more challenging. We explain how to handle this interference situation with different eICIC techniques in Sections 5.3–5.5. In the uplink, the interference situation is the same as that in the macro–pico deployment scenario, which can be efficiently solved by the power control techniques in Release 8/9.

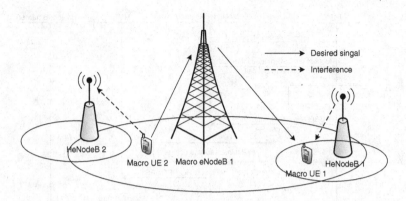

Figure 5.7 The interference situations in the macro–femto deployment scenario

5.3   Time Domain Techniques

From the discussion in the previous section, there is a potential need of interference management in the downlink to avoid severe interference caused by macrocells in the macro–pico
deployment scenario, and by non-accessible CSG femtocells in the macro–femto deployment
scenario. The LTE Release 8/9 ICIC techniques work well for PDSCH interference management by frequency domain scheduling and resource partitioning. However, it is difficult to
apply these techniques to the control channels and reference signals as they have fixed locations in the frequency domain. For example, the PDCCH spans the entire bandwidth in the first
few OFDM symbols in each subframe. Another example is the PSS/SSS, which are always
transmitted in the central 6 RBs of the system bandwidth. The need for interference mitigation
of those control channels and reference signals is the main motivation of the eICIC techniques
that were introduced in Release 10.

In this section, we focus on the time domain eICIC techniques. We first introduce Almost
Blank Subframe (ABS) in Section 5.3.1, which is the key concept for the time domain interference mitigation. We then discuss how to utilize ABSs for interference management in
the macro–pico and macro–femto deployment scenarios in Section 5.3.2. Afterwards the
UE measurement and reporting procedures for a network configuring ABSs are discussed in
Section 5.3.3. The required backhaul support is discussed in Section 5.3.4. Finally, we provide some system-level simulation results to show the performance gain of the time domain
techniques in Section 5.3.5.

5.3.1   Almost Blank Subframe

The basic idea of the time domain techniques is to mute certain subframes of some cells in order
to reduce the interference to the other cells [5, 6, 7]. We call a cell that causes interference to
UEs of another cell an *aggressor cell* and the latter the *victim cell*; the UEs that are interfered
are referred to as the *victim UEs*. Ideally, the muted subframes configured by an aggressor
cell should be totally blank (i.e. all REs are muted) in order to reduce the interference as
much as possible. However, the broadcast channel (i.e. the PBCH) and the reference signals
(e.g. the CRS and the PSS/SSS) cannot be muted as they are needed to support important

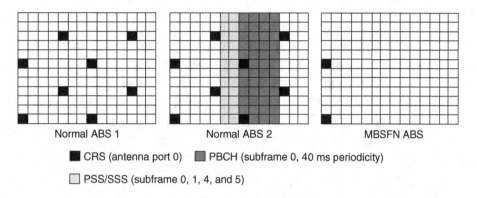

Normal ABS 1   Normal ABS 2   MBSFN ABS

■ CRS (antenna port 0)   ■ PBCH (subframe 0, 40 ms periodicity)

☐ PSS/SSS (subframe 0, 1, 4, and 5)

Figure 5.8   Normal and MBSFN ABSs

legacy operations such as cell acquisition and maintaining link connectivity. In addition, the PDCCH may have to be transmitted to support an uplink transmission such as uplink grants. Based on the above considerations, ABSs that are introduced in Release 10 allow only unicast PDSCH transmissions to be suspended. Interference is therefore reduced significantly but not completely.

ABSs can be configured by a cell in either normal subframes or MBSFN subframes. If a normal subframe is configured as an ABS, the CRS and the PBCH (where applicable) are still transmitted as shown in Figure 5.8. If a MBSFN subframe is configured as an ABS, it does not contain the CRS in the PDSCH region and thus causes less interference than a normal ABS. Note that the PDCCH region is also almost blank in ABSs since no DCI corresponding to the unicast PDSCH transmission is transmitted. As mentioned earlier however, if an uplink grant is scheduled in that subframe there could be a corresponding PDCCH transmission in some REs.

Given the fact that a MBSFN ABS causes less interference than a normal ABS, the obvious question is whether it is possible to configure all ABSs in MBSFN subframes. Unfortunately, this is not possible as it may violate the uplink HARQ process configurations. To understand this, consider a simple example in the macro–pico deployment scenario as shown in Figure 5.9, where a macrocell configures ABSs to reduce the interference to a picocell. If the picocell transmits a PUSCH grant for a UE in subframe 1 (which is configured as MBSFN ABS by the macrocell), it will transmit the HARQ ACK/NACK for the scheduled PUSCH transmission in subframe 9. In this case, it is desirable for the macrocell to configure subframe 9 as a normal ABS to protect the HARQ ACK/NACK transmission.

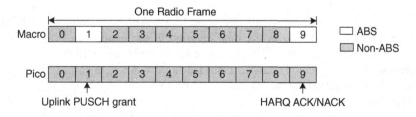

Figure 5.9   An example of using ABSs to protect uplink grants and HARQ ACKs/NACKs

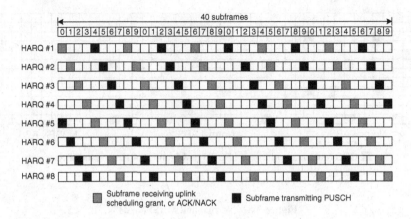

Figure 5.10   Uplink HARQ timing in FDD

Subframe 9 cannot be configured as a MBSFN subframe, however. Only subframes 1, 2, 3, 6, 7, and 8 can be configured as MBSFN subframes. This means a cell cannot configure all its ABSs in MBSFN subframes, but has to configure some ABSs in normal subframes also.

The above example reveals an important design principle of ABS patterns, i.e. it is desirable to align the ABS pattern with the uplink HARQ timing. In uplink, synchronized HARQ is adopted to reduce control signaling overhead and UE complexity. The HARQ timing pattern in FDD mode supports up to 8 HARQ processes and is illustrated in Figure 5.10. We see that the HARQ timing pattern has a periodicity of 40 ms. Based on the above design principle, we can have the following 8 basic ABS patterns with the same periodicity of 40 ms, each of them aligned with one of the HARQ processes, where 1 indicates that the corresponding subframe is configured as an ABS:

- ABS pattern 1: 10000000100000001000000010000000,
- ABS pattern 2: 01000000010000000100000001000000,
- ABS pattern 3: 00100000001000000010000000100000,
- ABS pattern 4: 00010000000100000001000000010000,
- ABS pattern 5: 00001000000010000000100000001000,
- ABS pattern 6: 00000100000001000000010000000100,
- ABS pattern 7: 00000010000000100000001000000010,
- ABS pattern 8: 00000001000000010000000100000001.

The above basic ABS patterns have the same 12.5% ABS ratio, which is defined as the percentage of ABS subframes in the total duration of the pattern. If a higher ABS ratio is desirable, two or more basic ABS patterns can be combined together. For example, the combination of ABS patterns 1 and 2 results in a new ABS pattern 1100000011000000 1100000011000000, which has 25% ABS ratio and is still aligned with the uplink HARQ timing.

In TDD mode, the number of uplink HARQ processes and their timing patterns depend on the UL/DL configuration. Figure 5.11 illustrates an example of the HARQ timing in TDD

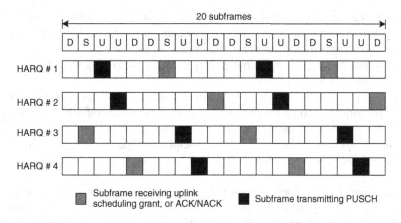

Figure 5.11   Uplink HARQ timing in TDD configuration 1

configuration 1, where up to 4 HARQ processes are supported[2]. Based on the same design principle, we can have the following 4 basic ABS patterns with 20 ms periodicity@

- ABS pattern 1: 00000010000000001000,
- ABS pattern 2: 00000000010000000001,
- ABS pattern 3: 01000000000100000000,
- ABS pattern 4: 00001000000000100000.

The above basic ABS patterns have the same 10% ABS ratio. Similar to the FDD case, a new ABS pattern with a higher ABS ratio can be generated by combining multiple basic ABS patterns.

5.3.2   ABS Use Cases

In this section, we explain how to utilize ABSs in heterogeneous networks. We consider the following two typical deployment scenarios:

1. the macro–pico deployment scenario, where the pico UEs in the CRE region of a picocell suffer strong interference from a macrocell; and
2. the macro–femto deployment scenario, where the macro UEs that are located in the proximity of a CSG femtocell suffer strong interference from it.

In the macro–pico deployment scenario, the expected behaviors of the macro- and picocells are described in the following.

Step 1: The macrocell predicts that the traffic demand may exceed its capacity and it has to offload some UEs to the picocell.

---

[2] This is based on the assumption of no Transmission Time Interval (TTI) bundling.

Step 2: The macrocell hands over some UEs to the picocell by employing CRE.

Step 3: The macrocell mutes some subframes by configuring ABSs. These subframes are also called *protected subframes* from the context of the picocell.

Step 4: The macrocell informs the picocell of the ABS pattern via the X2 interface (this is explained in Section 5.3.4).

Step 5: The picocell schedules those UEs in the protected subframes based on the received ABS pattern. It configures its UEs to report channel state information of the protected and non-protected subframes separately as the nature of interference is very different in the two types of subframes. This principle is called restricted measurements and is explained in Section 5.3.3.

The percentage of ABSs configured by the macrocell depends on the CRE bias. In principle, if a large CRE bias is configured, more UEs will be served by the picocell and thus a higher ratio of ABSs is desirable. The CRE bias and ABS ratio need to be optimized jointly to achieve the maximum system throughput.

An example of employing ABSs in the macro–pico deployment scenario is shown in Figure 5.12. In this example, 40% ABS ratio is configured by the macrocell. Specifically, subframes 1, 3, 5, 7, and 9 are configured as ABSs, of which subframes 1, 3, and 7 are configured as MBSFN ABSs to further reduce the interference to the picocell. Based on the above ABS configuration, the picocell will schedule its cell-edge UEs in the protected subframes (i.e. subframes 0, 2, 4, 6, and 8) and use the remaining unprotected subframes for cell-center UEs.

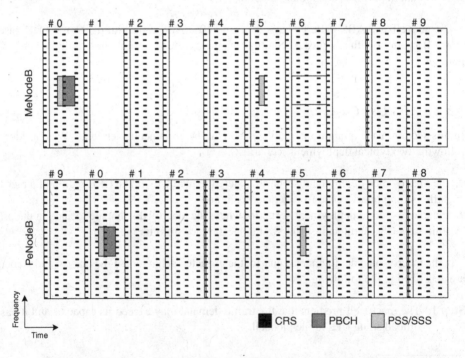

Figure 5.12   An example of employing ABSs in the macro–pico scenario

Note that the subframes of the two eNodeBs are *shifted*, i.e. subframe 0 of the picocell is aligned with subframe 1 of the macro. Subframe shifting has been introduced in LTE to ensure that the PSS/SSS and the PBCH transmitted by the picocell can be protected. With subframe shifting, the macro transmits PSS/SSS signals and PBCH channels in a different subframe than the pico. The macro has two ways to avoid interference to pico PSS/SSS and PBCH when subframe shifting is used, described in the following:

- the macro configures an ABS corresponding to the subframe where pico transmits PSS/SSS and PBCH; and
- the macro configures a non-subframe ABS corresponding to the subframe where pico transmits PSS/SSS and PBCH, but it avoids scheduling any UEs in the central 6 RBs of the subframe (because the pico PSS/SSS and PBCH will be transmitted in the central 6 RBs).

Another method called OFDM symbol shifting can be adopted to prevent interference. In this method, the OFDM symbols of the two cells are shifted (instead of subframes as in subframe shifting). OFDM symbol shifting can protect the PDCCH transmitted by the victim cell in the protected subframes. Without OFDM symbol shifting, the PDCCH regions of the two cells collide at least in the first OFDM symbol. As a result, the PDCCH transmitted by the victim cell suffers from the CRS interference from the aggressor cell in each subframe, regardless of whether it is configured as a ABS or not. With OFDM symbol shifting, the PDCCH transmitted by the victim cell could experience zero interference in the protected subframes.

In the macro–femto deployment scenario, the femtocell is an aggressor cell and the macrocell is a victim cell. ABSs can be configured by the femtocell to reduce the interference to the macro UEs that are close to it. The expected behaviors of the macro- and femtocells are as follows.

Step 1: The femtocell detects some macro UEs in its proximity.
Step 2: The femtocell mutes some subframes by configuring ABSs.
Step 3: The femtocell informs the macrocell of the ABS pattern via the Operations, Administration, and Maintenance (OAM) interface. OAM is the collective name given to the processes, activities, tools, and standards involved with operating, administering, managing, and maintaining any system. These fall under the core network functionalities in LTE.
Step 4: The macrocell schedules the victim UEs in the protected subframes based on the received ABS pattern. At the same time, it configures various restricted measurements for its associated UE (as explained in Section 5.3.3).

The percentage of ABSs configured by the femtocell again depends on the data traffic of the victim macro UEs (that are located in the dead zone around the HeNodeB) as well as the data traffic and location of the femtocell. In principle, a higher ABS ratio is desirable if the number of the victim macro UEs is larger and the femtocell is deployed on the edge of the macrocell. The ABS ratio needs to be optimized in order to achieve a good throughput balance between the macro and femto UEs.

An example of employing ABSs in the macro–femto deployment scenario is shown in Figure 5.13. In this example, 20% ABS ratio is configured by the femtocell assuming that the number of victim macro UEs is relatively small. Specifically, subframes 1 and 9 are configured

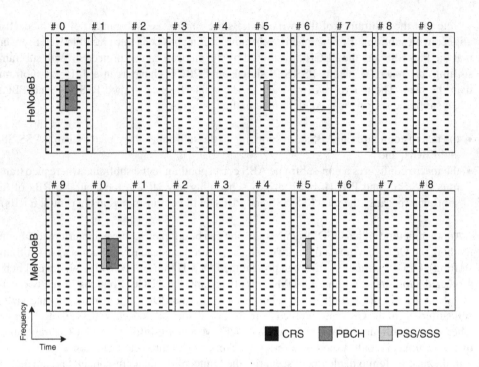

Figure 5.13 An example of employing ABSs in the macro–femto deployment scenario

as ABSs, of which subframe 1 is configured as a MBSFN ABS to further reduce the interference to the macrocell. According to such a ABS configuration, the macrocell will schedule the UEs that are close to the HeNodeB in the protected subframes (i.e. subframes 0 and 8) and use the remaining unprotected subframes for the other associated UEs.

Subframe shifting is used in this example to protect the PSS/SSS and PBCH transmitted by the eNodeBs. OFDM symbol shifting, mentioned in the macro–pico case, can also be used.

From the above two examples, we see that the victim cell needs to know the ABS pattern, i.e. which subframes are configured as ABSs, of the aggressor cell in order to schedule the victim UEs in those subframes. In Section 5.3.4, we explain how the ABS pattern is exchanged between the aggressor cell and the victim cell via the backhaul link.

5.3.3 UE Measurement and Reporting

In a LTE network, a UE performs various measurements for different purposes (e.g. cell selection/reselection, handover, or link adaptation) and reports them to its serving cell periodically. In this section, we discuss the impact of configuring ABSs on the following three UE measurement and report procedures:

- Radio Resource Management (RRM) measurements,
- Radio Link Monitoring (RLM) measurements, and
- Channel State Information (CSI) measurements.

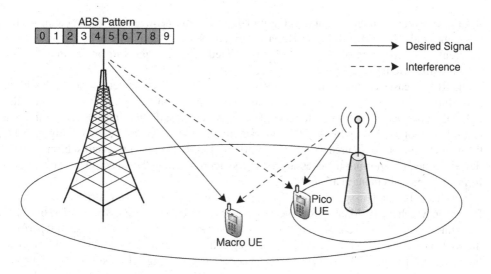

Figure 5.14 UE measurement and report in the macro–pico deployment scenario

Before going into the details of the above UE measurements, we first give an example to show how the configuration of ABSs affects UE interference measurements. Consider a typical macro–pico deployment scenario as shown in Figure 5.14, where the macrocell employs ABSs to reduce interference to the picocell. Obviously, the interference characteristics experienced by a pico UE in the protected subframes (i.e. subframes 1, 3, and 9) is different from that in the unprotected subframes. If the pico UE does not take such differences into account, it will perform interference averaging over some protected subframes together with some unprotected subframes, which leads to very inaccurate interference estimation.

To solve the above-mentioned issue, the concept of restricted measurements was introduced in Release 10. This means that a UE can be configured to perform a certain measurement over a restricted set of subframes. In the interference measurement example, the subframes could be divided into two sets of subframes: one for the protected subframes and the other one for the unprotected subframes. The pico UE can perform independent interference measurement and estimation of these two measurement sets. Note that the configuration of ABSs does not affect the interference measurement of the macro UEs, since the interference characteristics from the picocell does not vary across subframes.

The RRM measurements are used for cell selection/reselection and handover for the RRC_IDLE and RRC_CONNECTED UEs, respectively. These measurements include Reference Signal Received Power (RSRP) and Reference Signal Received Quality (RSRQ). The RSRP measurement is performed based on CRS, which does not depend strongly on whether it is taken in ABS subframes or non-ABS subframes. The RSRP measurement therefore remains unchanged in Release 10. The RSRQ is defined as the ratio between the RSRP and Received Signal Strength Indicator (RSSI), which comprises the linear average of the total received power (including the power of the desired signal power, interference, and noise) observed only in the OFDM symbols containing CRS for antenna port 0. Since the RSSI value varies depending on whether the measurement is taken in ABSs or non-ABSs, the RSRQ measurement needs to be be restricted to a limited set of subframes. In the example shown in Figure 5.14, the macro UE can be configured to perform the RSRQ measurement in all subframes while the pico UE can be configured to perform the measurement in the protected subframes only. Note that different RRM restricted measurement sets can be configured for different UEs depending on whether their associated cells are aggressor or victim cells, and their proximity to interfering neighboring cells.

The RLM measurements are used for a RRC_CONNECTED UE to determine the downlink radio link quality of the serving cell to determine if synchronization has been lost or radio link failure has occurred. Similar to the RSRQ measurements, the serving cell can configure a restricted set of subframes for the RLM measurements. In the example shown in Figure 5.14, the pico UE can be configured to perform RLM measurement in the protected subframes only in order to avoid reporting unnecessary loss of synchronization. The macro UE can still perform RLM measurement in all subframes.

In Release 10, the UE-specific RRC signaling is used to indicate the restricted sets of subframes on which the RLM and RRM measurements are to be performed. Specifically, the RRC signaling can configure a common restricted set of subframes on which the UE should perform the RLM and RRM measurements on the serving cell. Another restricted set of subframes can be configured for the RRM measurements on certain neighboring cells specified by PCIs. For the neighboring cells whose PCIs are not listed, the RRM measurement is performed in all subframes.

In order to support MIMO operation, the UE can be configured to perform CSI measurements and report Channel Quality Indicator (CQI), Precoding Matrix Indicator (PMI), and Rank Indicator (RI). The CSI measurements, especially the CQI report, provide the serving cell with information related to instantaneous SINR, which is used for UE scheduling and link adaptation. If ABSs are configured by a cell, the SINR experienced by its victim UEs (which are associated with other cells) will dramatically change from the protected subframes to the unprotected subframes. The CSI measurements for those UEs should therefore be taken in a restricted set of subframes. In the example shown in Figure 5.14, the pico UE can be configured to measure the CSI in protected subframes only. For the macro UE, it can still measure CSI in all subframes.

In Release 10, the UE-specific RRC signaling can configure two independent and non-overlapping restricted sets of subframes for a UE to perform CSI measurements. This allows the serving cell to compare the UE's CQI reports for different sets of subframes, e.g. ABSs and non-ABSs. If the serving cell has two dominant interfering cells configured with different ABS patterns, the first set of subframes can include the common ABSs configured by both interfering cells, and the second set could contain only the ABSs configured by one of the interfering cells.

Consider a general scenario with an aggressor cell and a victim cell, where the aggressor cell configures ABSs to reduce interference to the victim cell. From the above discussion, the general principle of configuring restricted subframe sets for RRM, RLM, and CSI measurements is that the restricted measurement sets need to be configured for the UEs in the victim cell based on the ABS pattern of the aggressor cell. The UEs in the aggressor cell do not usually have restrictions for those measurements.

5.3.4 Backhaul Support

The ABS-based time domain techniques need backhaul support to ensure their effectiveness. If ABSs are configured by an aggressor cell, the ABS pattern needs to be shared with its neighboring victim cells which would benefit from this information. With the knowledge of the ABS pattern, the victim cell can schedule the cell-edge UEs in the protected subframes. In addition, it could configure various restricted measurement sets for its associated UE as discussed in the previous section.

In the macro–pico deployment scenario, the ABS pattern configured by a macrocell is indicated to its neighbor cells via the X2 interface. Specifically, two *ABS bitmaps* are signaled from a cell (typically a macrocell) to another cell (typically a picocell). Each bit of the first bitmap corresponds to a subframe and indicates whether the subframe is configured as an ABS or not. The second ABS bitmap represents the restricted measurement set, which is a subset of the first bitmap. It is intended to inform the receiving cell about the appropriate subframe sets for RRM, RLM, and CSI measurements.

The ABS bitmaps are transmitted as part of the *Load Indication* and *Resource Status Reporting Indication* in the X2 application protocol. They can be updated semi-statically, i.e. no faster than the Release 8 frequency-domain RNTP update. In order to align the ABS pattern with the time of uplink HARQ processes, as discussed in Section 5.3.1, the ABS bitmaps have different periodicities for different FDD/TDD configurations. These are summarized as follows:

- 40 ms for FDD mode,
- 70 ms for TDD configuration 0,
- 20 ms for TDD configuration 1–5, and
- 60 ms for TDD configuration 6.

A cell can request another cell to configure ABSs by sending an *Invoke Indication* via the X2 interface. This happens when the cell identifies the other cell as a dominant interferer. The request initiates the ABS configuration process of the cell receiving the request.

A cell which receives the ABS bitmaps from another cell can return two messages: a *Usable ABS Pattern* message and an *ABS Status* message. The Usable ABS Pattern message indicates which ABSs are used by the cell to schedule its associated cell-edge UEs. The ABS Status message indicates the percentage of the RBs scheduled for data transmissions in the ABSs indicated by the Usable ABS pattern. The Usable ABS Pattern and ABS Status messages help the other cell to adjust its configured ABS ratio and pattern.

Figure 5.15 depicts a typical message exchange via the X2 interface between a macrocell and a picocell for ABS coordination. The picocell identifies the macrocell as one of the dominant interferers and sends an *Invoke Indication* to request it to configure ABSs. The macrocell then configures an ABS pattern and informs the picocell about this pattern by sending the ABS bitmaps. The picocell schedules its associated UEs based on the received ABS pattern. It returns the *Usable ABS Pattern* together with *ABS Status* messages to help the macrocell to determine whether the number of configured ABSs should be increased or reduced.

5.3.5 Simulation Results

In this section, we provide some simulation results to show how the CRE and ABS configurations affect the system performance in terms of traffic offloading and system throughput. In particular, we look into the following aspects:

- how the picocells (without or with CRE) help the network to offload data traffic;
- how the overall throughput performance benefits from the employment of the CRE and ABSs; and
- how the CRE affects the ABS configuration and overall throughput performance.

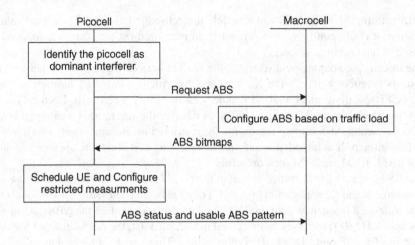

Figure 5.15 ABS-related message exchanged over the X2 interface

Table 5.1 Simulation assumptions

Parameter	Configuration
System bandwidth	10 MHz
Central frequency	2 GHz
Traffic model	Full buffer
Channel model	ITU UMa for macro, UMi for pico

Consider a macro–pico deployment according to the models described in [8]. In the simulation, 57 macrocells are deployed following a hexagonal layout with 500 m Inter-Site Distance (ISD). For each macrocell, a fixed number of picocells and 25 UEs are deployed randomly within the coverage area. The transmit powers of the macrocell and the picocell are assumed to be 46 dBm and 30 dBm, respectively. Other important simulation assumptions are listed in Table 5.1.

Figure 5.16 shows the percentage of UEs served by picocells when the number of picocells per macrocell is varied. Note that 0 dB CRE is assumed here. It is seen that more UEs will be served by the picocells if there are more picocells per macrocell. However, the rate of growth in number of associated UEs to the pico slows down as the picocells increase. This is because, as the number of picocells increase, two of them may be deployed close to each other; they therefore end up targeting the same set of UEs to offload from the macro. In other words, the CRE regions of two picocells have a higher chance of being overlapped if there are more picos per macro area.

In order to show the impact of the CRE on network offloading, Figure 5.17 depicts the percentage of UEs served by picocells as the CRE bias changes. It is seen that the curve for a higher number of picos/macro (e.g. 10) is *flatter* than that for lower number of picos/macro (e.g. 1). If there are many picos deployed in the macro area, then the ABS performance is less

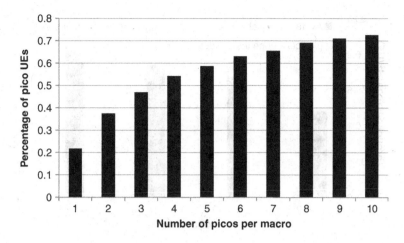

Figure 5.16   Offloading results as the number of picocells in a macro area is increased

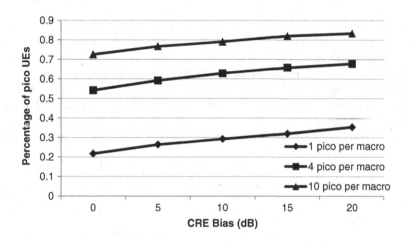

Figure 5.17   The percentage of pico UEs with different CRE bias values

sensitive to changes in CRE bias. This is because the CRE regions of picocells may overlap in the dense deployment scenario (which was also observed in Figure 5.16).

When a macrocell configures ABSs, the ABS ratio is a key parameter determining the effectiveness of ABS and system performance. Figure 5.18 shows the system performance with different ABS ratios in terms of the cell-average throughput, the cell-edge throughput for macro UEs, and the cell-edge throughput for pico UEs [9]. In the simulation, 2 picocells per macrocell and 15 dB CRE are assumed. It is seen that a high ABS ratio is beneficial for the UEs on the edge of the picocells but reduces the throughput of macro UEs. The reason is that a large number of ABSs configured by a macrocell reduces interference to the neighboring picocells at the cost of a reduced number of subframes for the transmission of its own UEs.

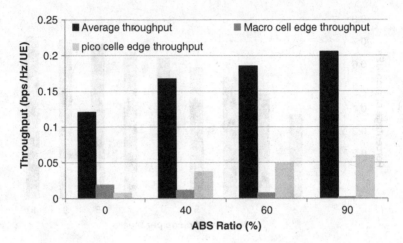

Figure 5.18 The throughput performance with different ABS ratios. (Source: Texas Instruments 2010 [9]. Reproduced with permission of ETSI.)

Note that the quantitative observations on cell-average throughput depends on the specific simulation assumptions such as CRE bias, scheduler, and link adaptation mechanisms. In general, the optimal ABS ratio to maximize the cell-average throughput is determined by many factors, for example, the number of picocells per macrocell and the CRE bias.

The CRE bias determines the amount of UEs offloaded to picocells in a given macro–pico deployment scenario. In order to show the benefit of CRE, the cell-average and 5% worst UE throughput results with different CRE bias values are provided in Figure 5.19 and Figure 5.20 respectively. The throughput results of non-ABS without CRE (shown as dashed lines) is considered as the base line for comparison purpose.

We see that the ABS increases the overall throughput of all UEs over no ABS for all CRE bias values. The gain reduces with increasing CRE bias. The 5% worse UE throughput is worse

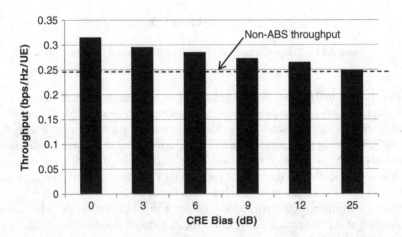

Figure 5.19 The cell-average throughput for all UEs with different CRE bias values. (Source: LG Electronics 2010 [10]. Reproduced with permission of ETSI.)

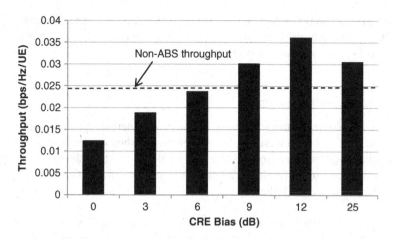

Figure 5.20 The cell-edge throughput with different CRE bias values. (Source: LG Electronics 2010 [10]. Reproduced with permission of ETSI.)

when ABS is utilized if the CRE bias is relatively small (less than 6 dB). In this case, the 5% worse UEs are actually macro UEs whose throughput is decreased compared with non-ABS due to the reduced resource of macrocells. When the CRE bias becomes larger, more UEs will be offloaded to picocells. The remaining macro UEs will have a greater chance of being scheduled by macrocells and thus the cell-edge throughput is better than non-ABS. The results suggest choosing a medium value of CRE bias such as 6–9 dB.

5.4 Power Control Techniques

Transmit power control is one of the more simple but effective ways for a cell to manage the interference to its neighboring cells. Technically, the ABS-based eICIC discussed in the previous section is a special power control scheme: a cell transmits with zero power in ABSs and with normal power in non-ABSs. In this section, we discuss more general power control schemes for downlink interference management. The target scenario to use power control for interference management is first discussed in Section 5.4.1. Next, two power control schemes introduced in Release 10 are explained in Section 5.4.2. Simulation results are provided in Section 5.4.3 to illustrate the performance gain of these two schemes.

5.4.1 Target Scenario

In the macro–pico deployment scenario, the macrocell can potentially use power control for interference management to the pico UEs in the CRE region. However, it is not desirable to reduce the transmit power of a macrocell since it will reduce the macro coverage and increase UE outage probability.

In the macro–femto deployment scenario, a CSG femtocell may cause strong interference to the macro UEs that are close to it but do not have the CSG access. It is possible for the femtocell to perform autonomous power control to reduce the interference to these victim UEs. Note that the reduced transmit power of the femtocell will also reduce its coverage. However, a femto UE in the CRE region can always switch to the macro if its connection to the femtocell becomes weak. This is the main scenario of interest for power control.

5.4.2   Power Control Schemes

In the macro–femto deployment scenario, no direct X2 interface exists to support real-time coordination. The femtocell therefore has to determine when to perform power control by itself. An autonomous mechanism has been suggested [11] in which the femtocell triggers the power control process by detecting the existence of victim macro UEs in its vicinity. The basic procedure is summarized as follows.

Step 1: The femtocell measures the downlink RSRP from its neighboring macro base stations. This is called the *Network Listen Mode*, where the femtocell essentially acts as a UE and can decode certain DL transmissions from the macro.

Step 2: If one of the RSRP values is larger than a pre-defined threshold, the femtocell will trigger power control for interference management. Otherwise, it transmits with maximum power. The intuition is that a large DL RSRP from the macro indicates that the macro BS is close to the femtocell. There is therefore a higher chance that some macro UEs are located close to the femtocell and become affected by interference.

Note that the downlink RSRP measured by the femtocell depends on its location. The threshold that a femtocell sets to trigger power control for interference management should therefore also depend on the location of the femtocell. For example, a higher value is expected for the femtocell located on the macrocell edge.

The power control schemes for the interference coordination in the macro–femto deployment are documented in [11]. In particular, the following two power control schemes are suggested.

Scheme 1: Power control based on the interference measurement from the dominant macrocell.

Scheme 2: Power control based on the path loss between the HeNodeB (femtocell) and the victim macro UE.

In the first scheme, the HeNodeB measures the RSRP from the most dominant macrocell and adjusts the transmit power as follows:

$$P_{tx} = \text{median}(\alpha P_m + \beta, P_{max}, P_{min}), \tag{5.1}$$

where P_{max} and P_{min} are the predefined maximum and minimum transmit powers, respectively, and P_m denotes the RSRP from the dominant macrocell. The intuition is that, if the received power from the macro at the femto is high, the femto can assume that the received power from the macro at the macro UEs would also be high with high probability. Under this assumption, the femtocell can raise its power without causing significant degradation in SINR for the macro UEs. The femto would want to increase its transmit power as much as possible without degrading the macro UEs SINR in order to guarantee a good SINR for its own UEs. The parameter

α allows the slope of the power control mapping curve to be altered and the parameter β can be used to alter the range of the power control. In the case where the HeNodeB is unable to detect any macrocells, it is free to use maximum transmit power.

The drawback of the first scheme is that it uses the macro base station to femto link quality as a measure for the macro UE to the femto base station quality. In the second power control scheme, the femtocell tries to estimate the path loss between itself and the macro UE and adjusts its transmit power based on this estimated value. Specifically, the femtocell adjusts its transmit power as

$$P_{tx} = \text{median}(P_m + P_{offset}, P_{max}, P_{min}), \tag{5.2}$$

where P_{offset} is a power offset representative of the path loss between the HeNodeB and the macro UE. This is estimated as

$$P_{offset} = \text{median}(P\_offset\_o + K * LE, P\_offset\_max, P\_offset\_min), \tag{5.3}$$

where P_offset_o is a predetermined power offset value corresponding to the indoor path loss; K is an adjustable positive factor which can be determined based on the priority of HeNB operation; LE is the estimated penetration loss (with a typical value between 10 and 30 dB); and $P\_offset\_max/P\_offset\_min$ is the maximum/minimum value of the P_offset.

The path loss between the HeNodeB and the macro UE is estimated based on the difference between the estimated UL transmit power and the UE received power at the femtocell which the femto can measure. The femto estimates the UL transmit power of the UE based knowledge about the UE power control parameters and the macro DL RSRP measured at the femto, which it assumes to be the same as the macro DL RSRP at the macro UE.

5.4.3   Results from Realistic Deployments

In this section performance of the power control schemes described in the previous section are provided for realistic deployments mandated by 3GPP. Consider a macro–femto deployment according to the models described in [8]. In the system, 57 macrocells are deployed following a hexagonal layout with 500 m ISD. In each macrocell, one cluster of femtocells is deployed randomly. For each macrocell, 25 UEs are dropped uniformly within the coverage area, some of which are macro UEs that cannot access the femtocell.

The femtocell cluster is modeled by a dual stripe model as shown in Figure 5.21. In the dual stripe model, two buildings are separated by an open strip, each of which consists of 2 by 10 apartments of size 10 m × 10 m.

Figure 5.22 shows the PDCCH outage probability of macro and femto UEs with different power control schemes [12]. Two power control schemes are considered and compared with a baseline scheme without power control. The scheme denoted 'PC 1' is the first power control discussed in the previous section with $\alpha = 1$ and $\beta = 70$. The scheme denoted 'PC 2' is the second power control scheme. We see that there is a clear trade-off between the Macro UE outage and the femto UE outage. Power control scheme 2 is slightly better than scheme 1 due to the extra knowledge of path loss between the femtocell and the victim macro UE. Figure 5.23 shows the PDCCH outage probability of the indoor and outdoor macro UEs with the two power control schemes. Compared with the baseline scheme, the power control schemes significantly reduce the outage probability of the indoor macro UEs. The outdoor macro UEs suffer significantly less interference from the femtocell than the indoor macro UEs, even without power

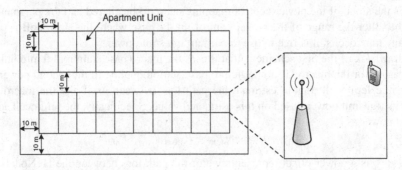

Figure 5.21   The dual trip model for femtocell modeling

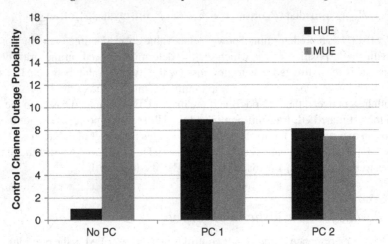

Figure 5.22   The PDCCH outage probability of macro and femto UEs with different power control schemes. (Source: Alcatel-Lucent 2010 [9]. Reproduced with permission of ETSI.)

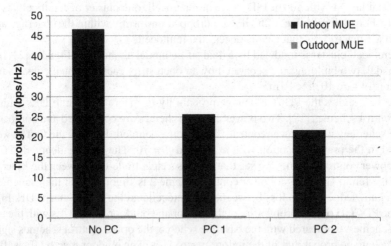

Figure 5.23   The PDCCH outage probability of indoor and outdoor macro UEs with different power control schemes. (Source: Alcatel-Lucent 2010 [9]. Reproduced with permission of ETSI.)

control, due to the wall penetration loss that significantly reduces the femto interference. The power control schemes do not therefore show much gain for outdoor macro UEs.

5.5   Carrier Aggregation-Based eICIC

Release 8/9 LTE systems can operate only on a single carrier with varying bandwidth from 1.4 MHz to 20 MHz. In order to further increase the system throughput, LTE-Advanced systems are expected to operate on a wider bandwidth (e.g. up to 100 MHz). One simple approach to extend the bandwidth, such that the single-carrier operations and protocols are still valid, is to aggregate multiple LTE component carriers (CCs) into a higher-bandwidth system. This is called *Carrier Aggregation* (CA), and was introduced in Release 10 [13]. However, multiple contiguous carriers are unlikely to be available in practice for aggregation. For example, the US operator AT&T only own four non-contiguous carriers as shown in Table 5.2. It is therefore desirable to allow UEs to receive data from multiple non-contiguous carriers simultaneously.

In Release 10 up to five CCs can be aggregated. The typical CA deployment scenarios for two carriers F1 and F2 (F1 < F2) are shown in Figure 5.24, which are summarized as follows.

- Scenario 1: F1 and F2 are frequencies in the same band and co-located in the same eNodeB. Both carriers provide similar coverage and mobility support.

Table 5.2   CA band owned by AT&T

Band index	Duplex mode	Frequency (MHz)
Band 2	FDD	1850–1910 (UL)/1930–1990 (DL)
Band 4	FDD	1710–1785 (UL)/2110–2155 (DL)
Band 5	FDD	824–849 (UL)/869–894 (DL)
Band 17	FDD	704–716 (UL)/734–746 (DL)

(Source: Shen *et al.* 2012 [13]. Reproduced with permission of IEEE.)

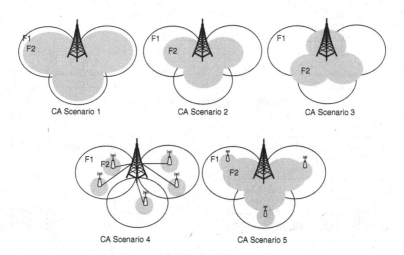

Figure 5.24   CA deployment scenarios [14]

- Scenario 2: F1 and F2 are frequencies in different bands and co-located in the same eNodeB. Only F1 provides sufficient coverage and mobility support while F2 is used for throughput enhancement.
- Scenario 3: F1 and F2 are frequencies in the same band and co-located in the same eNodeB. This is different from Scenario 2 as the main lobe of the antennas in F2 are directed towards the cell boundary of F1 to improve the cell-edge throughput. Only F1 provides sufficient coverage and mobility support.
- Scenario 4: F1 and F2 are frequencies in different bands. F1 is located in the macro eNodeB which provides sufficient coverage and mobility support. F2 is located in the Remote Radio Head (RRH) for throughput enhancement.
- Scenario 5: This is similar to Scenario 2 with the difference that frequency selective repeaters are deployed on F2 to extend its coverage.

The protocol structure of CA is shown in Figure 5.25. The CCs can be either contiguous or non-contiguous. From the user plane perspective, the CCs are transparent to the Packet Data Convergence Protocol (PDCP) and Radio Link Control (RLC) layers. The multiple CCs are just data transmission pipes managed by a single scheduler in the MAC layer. Each CC has its own Hybrid Automatic Repeat Request (HARQ) entity, maintaining a number of Release 8/9 compatible HARQ processes for the Physical layer.

From the control plane perspective, only the RRC_CONNECTED UEs can operate on multiple CCs via CA. CA does not impact on the procedures of the RRC_IDLE UEs. A CC is treated

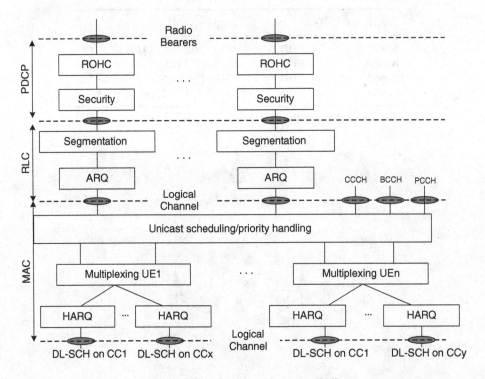

Figure 5.25 CA protocol structure [14]

Table 5.3 PCell versus SCell

Procedure	PCell	SCell
System information acquisition	Yes	No
Radio link monitoring	Yes	No
Cell deactivation	No	Yes

as a serving cell by the higher-layer procedures. Each UE has a single CC that provides all necessary control information such as RRC connection maintenance. This cell is defined as the Primary Cell (PCell). Other serving cells are defined as Secondary Cells (SCells). The main difference between a PCell and a SCell, summarized in Table 5.3, is explained as follows.

- A UE operating in the CA mode receives system information only from its PCell. The system information in a SCell is conveyed to the UE via dedicated RRC signaling.
- A UE operating in the CA mode only performs the RLM measurement on its PCell. This is intuitive since the important system and control information is sent by its PCell.
- A SCell can be activated or deactivated by RRC. For a deactivated SCell, the UE does not transmit/receive any uplink/downlink signal. For an activated SCell, the UE performs normal uplink transmission and downlink reception. A PCell always remains activated. Note that the network can perform SCell activation/deactivation for a UE to control the interference to its neighboring cells. This is explained in detail in Chapter 7. NAS mobility information from the MME is transmitted only on the PCell.

The configuration of PCell and SCell(s) is UE-specific. A RRC_IDLE UE establishes a RRC connection with its serving cell, which automatically becomes its PCell. Based on the CC on which the initial access is performed, different UEs may have different CCs as their PCell. When the RRC connection to the PCell is established, the network can further configure one or more CCs as SCells for the UE based on its traffic demand, the traffic load, and the interference at each CC. The system information of the SCells is signaled to the UE via dedicated RRC signaling. The network can also add, remove, or reconfigure SCells via dedicated RRC signaling. Note that the network can also change the PCell of a UE via handover procedure. From an operator perspective, it is beneficial to use low-frequency cells as PCells and high-frequency cells as SCells. This is because at lower frequencies the path losses are lower leading to higher coverage, which is desirable in PCells.

If the number of UL CCs is less than the number of DL CCs (as is normally the case), then a linkage between each DL CC to a UL CC has to be provided for purposes of transmission of UL control information related to a DL transmission (e.g. CSI feedback or HARQ transmission for the DL transmission). The linkage of the DL and UL CCs is signaled to UE via SIB-2 signaling.

Also recall that the PDCCH provides the control information for the transmissions of PDSCH. By default, each CC is configured to transmit the PDCCH for the PDSCH on the same CC. However, a PDCCH in a CC can carry control information about another CC's PDSCH. This is called *Cross-Carrier Scheduling*. The PDCCH that is transmitted to a UE therefore has to have the additional information about the CC in which the PDSCH lies. All the existing DCI formats carried by the PDCCH have been appended in Release 10 to include

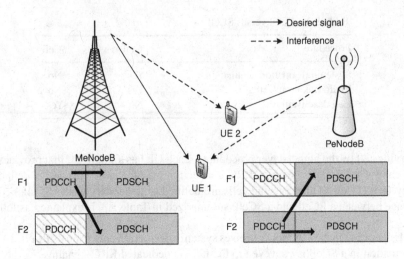

Figure 5.26 An example of PDCCH ICIC using cross-carrier scheduling

a 3-bit field called the Carrier Indicator Field (CIF). The value given in the CIF field indicates which CC has the PDSCH.

The main motivation of cross-carrier scheduling is to coordinate the inter-cell interference for the PDCCH in a multi-carrier deployment scenario. Consider a macro–pico deployment scenario as shown in Figure 5.26, where the MeNodeB and the PeNodeB share two carriers F1 and F2. In this case, the time-domain techniques can be used to coordinate the inter-cell interference on each CC individually, as explained in Section 5.3. However, an even better PDCCH interference coordination can be achieved if the cross-carrier scheduling is enabled, as the PDCCHs of different UEs are not in the same carriers and hence do not interfere.

With the cross-carrier scheduling, only one CC needs to be protected for the PDCCH transmission, which will be used by the eNodeB to schedule the PDSCH on any of its CCs. In the example shown in Figure 5.26, F1 and F2 are used by the MeNodeB and the PeNodeB for the PDCCH transmission, respectively. The PDCCH carried on either CC is used to schedule the PDSCH on both carriers. In this way, the two cells suffer zero interference from each other on the PDCCH region.

References

[1] Alcatel-Lucent and Alcatel-Lucent Shanghai Bell (2010) R1-105215, On the Potential Gains of Cell Range Expansion for the Macro-Pico Case. Alcatel-Lucent, Technical Report, October 2010.
[2] Texas Instruments (2010) R1-105294, On Cell Range Expansion. Texas Instruments, Technical Report, October 2010.
[3] Nokia Siemens Networks and Nokia (2010) R1-105552, Discussion of Pico Node Range Extension Benefits. Nokia, Technical Report, October 2010.
[4] CATT (2010) R1-102672, Evaluation of R8/9 Techniques and Enhancements for PUCCH Interference Coordination in Macro-Pico. CATT, Technical Report, May 2010.
[5] NTT DoCoMo (2010) R1-105724, Views on eICIC Schemes for Rel-10. NTT DoCoMo, Technical Report, October 2010.

[6] CATT (2010) R1-105183, Considerations on Time Domain Solution in Macro-Pico. CATT, Technical Report, October 2010.

[7] QualComm (2010) R1-105587, Details of Time-Domain Extension of Rel-8/9 Backhaul-Based ICIC. QualComm, Technical Report, October 2010.

[8] 3GPP (2010) Evolved Universal Terrestrial Radio Access (E-UTRA); Further Advancements for E-UTRA Physical Layer Aspects 3GPP TR 36.814, v9.0.0. Third Generation Partnership Project, Technical Report, March 2010.

[9] Texas Instruments (2010) R1-105903, Further Results on Cell Range Expansion. Texas Instruments, Technical Report, November 2010.

[10] LG Electronics (2010) R1-105354, Downlink Performance over Heterogeneous Networks. LG Electronics, Technical Report, October 2010.

[11] 3GPP (2012) Evolved Universal Terrestrial Radio Access (E-UTRA); FDD Home eNode B (HeNB) Radio Frequency (RF) Requirements Analysis 3GPP TR 36.921 v11.0.0. Third Generation Partnership Project, Technical Report, September 2012.

[12] Alcatel-Lucent, Alcatel-Lucent Shanghai Bell (2000) R1-104102, Power Control Techniques for Henb. Alcatel-Lucent, Technical Report, June 2010.

[13] Shen, Z., Papasakellariou, A., Montojo, J., Gerstenberger, D., and Xu, F. (2012) Overview of 3GPP LTE-Advanced carrier aggregation for 4G wireless communications. *IEEE Communications Magazine*, **50**(2), 122–130.

[14] 3GPP (2012) Evolved Universal Terrestrial Radio Access (E-UTRA) and Evolved Universal Terrestrial Radio Access (E-UTRAN), overall description 3GPP TS 36.300 v11.2.0. Third Generation Partnership Project, Technical Report, June 2012.

6

Release 11 Further Enhanced ICIC: Transceiver Processing

6.1 Introduction

In LTE downlink, the time domain techniques based on Almost Blank Subframes (ABSs) were adopted for enhanced Inter-Cell Interference Coordination (eICIC) in Release 10. For example, in a macro–pico deployment scenario, ABSs can be configured by a macrocell to significantly reduce the interference to its neighboring macro-/picocells. The residual interference from the macrocell in an ABS can however be significant as the Reference Signals (RSs) such as the CRS and Primary/Secondary Synchronization Signals (PSS/SSS) still have to be transmitted in the ABS for the purposes of backward compatibility. New techniques for interference mitigation in ABS are therefore needed to further improve system performance.

In this chapter, we discuss the Further enhanced Inter-Cell Interference Coordination (FeICIC) techniques considered in LTE-Advanced Release 11. We first introduce the target deployment scenario for FeICIC in Section 6.2, which is the same macro–pico deployment scenario as in Release 10, and highlight the issues that need to be addressed. We then explain the further enhancements of Release 10 eICIC for the macro–pico deployment scenario in Sections 6.3–6.5.

6.2 Typical Deployment Scenarios

LTE-Advanced Release 11 focuses on the same macro–pico deployment scenario that was considered in Release 10 in which picocells are deployed within the coverage area of macrocells for capacity/throughput enhancement. In this scenario, only a single carrier is available in the system. This is an important practical scenario that could be be present in many deployments. In the example shown in Figure 6.1, a UE served by the picocell (but located on the cell edge) suffers strong interference from its neighboring macrocell, especially when the Cell Range Expansion (CRE) is configured for that picocell. Release 10 time domain techniques can be employed in this example to reduce the interference from the macrocell. The macrocell configures some subframes as ABSs and, accordingly, the picocell schedules its cell-edge UEs in those protected subframes.

Heterogeneous Networks in LTE-Advanced, First Edition. Joydeep Acharya, Long Gao and Sudhanshu Gaur.
© 2014 John Wiley & Sons, Ltd. Published 2014 by John Wiley & Sons, Ltd.
Companion Wesite: www.ltehetnet.com.

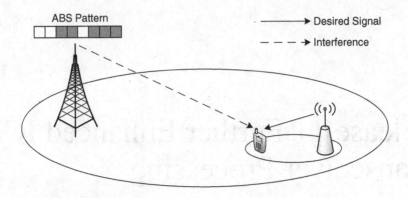

Figure 6.1 The single-carrier macro–pico deployment scenario

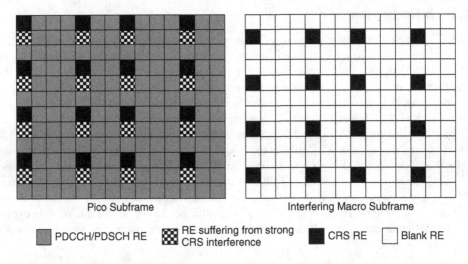

Figure 6.2 The CRS interference from the macrocell in the non-colliding CRS case

The above ABS-based ICIC solution has the following limitations, however.

- The pico UEs scheduled in a protected subframe may still suffer considerable interference from the macrocell [1]. In an ABS, up to 1/7 Resource Elements (REs) could be used by the macrocell for CRS transmission,[1] and thus the resulted interference cannot be ignored. The CRS interference could be even stronger if the pico CRE is configured or the CRS transmit power of the macrocell is boosted.[2] In Figure 6.2, we give an example to illustrate the impact of the CRS interference for the case where the CRS REs of the two cells are not aligned. This is called the *non-colliding CRS case*. In this example, a UE receives PDSCH from the picocell when the macrocell is transmitting an ABS. The PDSCH REs that do not collide

---

[1] This is based on the assumption that four antenna ports are configured by macro eNodeB.

[2] In a LTE system, up to 6 dB CRS power boosting can be used to enhance the macro coverage.

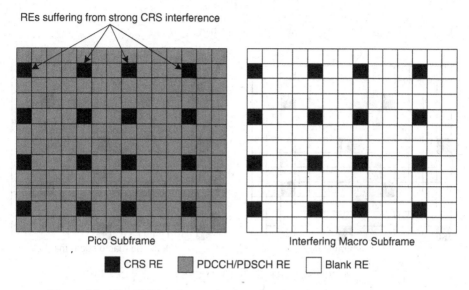

Figure 6.3   The CRS interference from the macrocell in the colliding CRS case

with the CRS transmitted from the macrocell suffer zero interference, while the PDSCH REs that collide with the CRS transmitted from the macrocell suffer strong interference. This non-uniformly distributed CRS interference will degrade the PDSCH demodulation performance of the protected pico UE.

- In the *colliding CRS case*, the macrocell and the picocell transmit CRS in the same REs as shown in Figure 6.3. In this case, the pico UE may fail to demodulate the CRS from its serving cell, even if the macrocell configures ABSs. This will affect the UE performance for the RLF and RRM measurements and also the PDSCH demodulation in CRS-based Transmission Modes. Note that the CRS interference can be reduced by configuring ABSs as Multicast-Broadcast Single-Frequency Network (MBSFN) ABSs, as they do not contain CRS in their PDSCH regions. However, not all ABSs are allowed to be configured as MBSFN ABSs (see Chapter 5).
- In an LTE/LTE-Advanced system, the PSS/SSS are transmitted on fixed time-frequency RE locations in all cells.[3] If the macrocell and the picocell have aligned radio frames and the same Cyclic Prefix (CP) configuration, the PSS/SSS/PBCH from the two cells will collide and interference with each other. An example of the PSS/SSS collision is as shown in Figure 6.4. Due to the high transmit power of the macrocell, it is difficult for the UE to decode the PSS/SSS/PBCH sent from the 'weak' picocell, even if the macrocell configures ABSs. In the case where a UE is handed over from the macrocell to the picocell with a large CRE bias, the UE may fail to detect the 'weak' picocell and thus cause handover failure.
- The macrocell is not allowed to schedule its associated UEs for the PDSCH transmission in ABSs.

Various enhancements for the ABS-based solution were proposed in Release 11 to address the above-mentioned limitations. These enhancements include the CRS interference

---

[3] In Release 12, the PSS/SSS may be transmitted on different time-frequency locations for New Carrier Type.

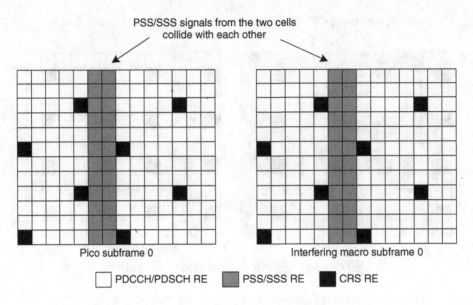

Figure 6.4   The PSS/SSS interference from the macrocell

cancellation, the PSS/SSS/PBCH interference cancellation for weak cell detection, and the non-zero-power ABS. We explain these enhancements in subsequent sections. In these scenarios, an eNodeB is assumed to be connected to its neighboring eNodeBs via an X2-based backhaul.

6.3   Techniques for Mitigating CRS Interference

In LTE-Advanced Release 11, various solutions were proposed for mitigating the CRS interference that occurs in ABS. Based on where the solutions are implemented, they can be divided into two categories: the Receiver-based (Rx-based) techniques and the Transmitter-based (Tx-based) techniques.

6.3.1   Receiver-Based Techniques

The Rx-based techniques are implemented at the UE to eliminate the CRS interference from its neighboring cells. Two Rx-based schemes have been adopted in Release 11 after a lengthy evaluation and discussion process. These are the CRS Interference Cancellation (CRS-IC) scheme [2, 3] and the PDSCH puncturing scheme [4, 3]. The first scheme can achieve better system performance while the latter has lower implementation complexity.

6.3.1.1   CRS-IC Scheme

In the CRS-IC scheme, the CRS interference from each dominant interfering cell is estimated and subtracted from the received signal sequentially, from the strongest to the weakest, as

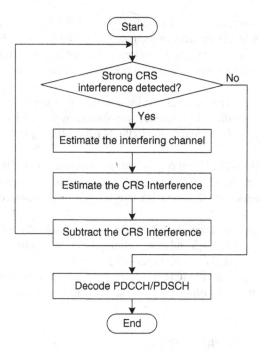

Figure 6.5   CRS interference cancellation at the UE receiver before decoding

shown in Figure 6.5. The UE first identifies the strongest interfering cell by measuring the RSRP levels of its neighboring cells. Note that the RSRP of the interfering cell can be higher than that of the serving cell. The estimate of the interfering channel is therefore expected to be accurate. After detecting the strongest interfering cell, the UE estimates the interfering channel by assuming that the CRS sequence transmitted by the interfering cell is known. After estimating the CRS interference from the interfering cell, the UE then subtracts it from the received signal. The above CRS-IC processing can be repeated to cancel the CRS interference from up to three dominant interferers.

The CRS-IC processing is a part of the MIMO receiver processing as shown in Figure 6.6. The MIMO receiver processing in Release 10 is usually an MMSE filter to separate the data

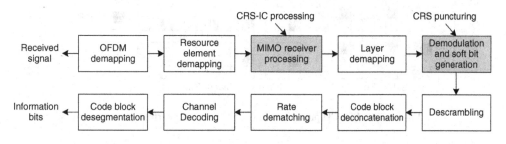

Figure 6.6   LTE receiver processing

streams from different spatial layers. In the Release 11 CRS-IC scheme, the UE performs the extra CRS-IC processing before feeding the received signal into the MMSE filter.

In order to facilitate the above CRS-IC operation, the serving cell has to provide the UE with the following information of each dominant interfering cell via higher layer signaling.

- Number of CRS ports: The UE uses this information together with the cell ID of the interfering cell[4] to generate the CRS sequence transmitted by the interferer. This information is also necessary for the UE to identify the PDSCH REs suffering CRS interference due to the interfering cell.
- MBSFN configuration: The interfering cell may have configured MBSFN subframes, which do not carry CRS in their PDSCH regions. This information is necessary for the UE to identify the subframes experiencing CRS interference from the interfering cell.

The CRS-IC receiver can cancel the CRS interference efficiently with the assistance of the serving cell in both colliding CRS and non-colliding CRS cases. However, the UE complexity is increased due to the extra CRS-IC processing, which reduces its battery life. Hence, in Release 11, a low-complexity PDSCH puncturing receiver has been proposed to handle the CRS interference for the non-colliding CRS case. This is discussed in the following section.

6.3.1.2 PDSCH Puncturing Scheme

In the PDSCH puncturing scheme, the UE detects the polluted REs that suffer strong CRS interference and simply discards them. In particular, after calculating the soft bits in the downlink receiver processing as shown in Figure 6.6, the UE punctures the soft bits carried by those polluted REs before feeding the bit sequence to the descrambler. In Figure 6.7, we provide a simple example of the PDSCH puncturing scheme. In this example, a UE receives the PDSCH from the picocell while the interfering macrocell transmits an ABS at the same time. The REs marked 'X' will be punctured to completely eliminate the CRS interference due to the macrocell.

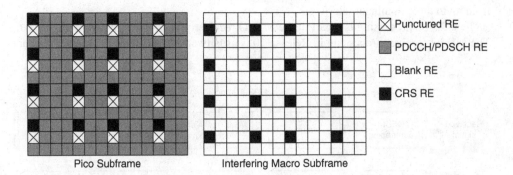

<table>
<tr><td></td><td>⊠ Punctured RE</td></tr>
<tr><td></td><td>▨ PDCCH/PDSCH RE</td></tr>
<tr><td></td><td>☐ Blank RE</td></tr>
<tr><td></td><td>■ CRS RE</td></tr>
</table>

Pico Subframe   Interfering Macro Subframe

Figure 6.7   PDSCH puncturing at the UE for CRS interference mitigation

---

[4] The UE obtains the cell ID of the interfering cell during the RSRP measurement process.

Note that the PDSCH puncturing scheme can only be used in the non-colliding CRS case. In the colliding CRS case, where the CRS transmissions of the serving cell and the interfering cell are in the same sets of REs, the polluted REs cannot be punctured.

From a signaling perspective, the PDSCH puncturing receiver requires the same signaling support from the serving cell as the CRS-IC receiver in order to identify the REs that need to be punctured.

The PDSCH puncturing scheme can completely remove the CRS interference from a particular interfering cell. Information needed for decoding is also lost during the puncturing process, however. There is a trade-off between the CRS interference reduction and the information lost. In practice, the UE usually punctures REs according to the CRS pattern of the strongest interfering cell. If it punctures more REs according to the CRS patterns of multiple interfering cells, a large number of information-bearing REs will be lost; this will cause performance degradation of the data reception.

The performance of the CRS-IC receiver, the PDSCH puncturing receiver, and the Release 10 receiver (without any CRS interference mitigation techniques) has been evaluated and compared [4]. The simulation results are shown in Table 6.1 and Table 6.2 for the cell-average throughput and the cell-edge throughput, respectively. In the evaluation, it is assumed that the CRS-IC receiver can completely cancel the CRS interference in ABSs. Such an ideal assumption provides the upper bound on the performance of various practical CRS-IC receivers. For the PDSCH puncturing receiver, it is assumed that the REs interfered by the CRS from the strongest interfering cell are punctured.

The evaluation results show that the Release 10 eICIC receiver experiences outage (i.e. zero throughput) on the cell edge when the CRE bias is greater than 12 dB. The ideal CRS-IC receiver and the PDSCH puncturing receiver can significantly increase the cell-edge throughput, especially for a large CRE bias. They also outperform the Release 10 Receiver in terms of the cell-average throughput. Furthermore, the PDSCH puncturing receiver can achieve similar

Table 6.1 Cell-average throughput (Mbps) of different types of UE receivers

CRE Bias (dB)	Release 10 Receiver	CRS-IC Receiver	PDSCH Puncturing Receiver
0	3.16	3.35	3.27
6	3.08	3.39	3.22
12	2.64	3.31	3.04
18	2.42	3.30	2.92

(Source: Hitachi 2011 [4]. Reproduced with permission of ETSI.)

Table 6.2 Cell-edge throughput (Mbps) of different type of UE receivers

CRE Bias (dB)	Release 10 Receiver	CRS-IC Receiver	PDSCH Puncturing Receiver
0	0.29	0.30	0.29
6	0.28	0.57	0.50
12	0	0.60	0.43
18	0	0.59	0.05

(Source: Hitachi 2011 [4]. Reproduced with permission of ETSI.)

Pico Subframe   Interfering Macro Subframe

Figure 6.8   CRS muting at the eNodeB

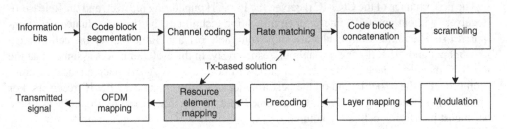

Figure 6.9   LTE transmitter processing

throughput performance as the ideal CRS-IC receiver for up to 12 dB CRE bias. In the case of 18 dB CRE bias, the PDSCH puncturing receiver is still beneficial for cell-edge UEs. However, the performance gap between the PDSCH puncturing receiver and the ideal CRS-IC receiver increases since the number of dominant CRS interferers increases with a high bias value (as UEs are located further away from the picocell), but the PDSCH puncturing receiver cannot handle more than one CRS interferer.

6.3.2   Transmitter-Based Techniques

The motivation of the Tx-based techniques is to save UE battery life by handling the CRS interference at the eNodeB. The basic idea is to avoid scheduling PDSCH on those REs that suffer from strong CRS interference. This way, no extra CRS interference suppression technique is needed at the UE receiver [5, 6, 7]. Consider the same example used for explaining the Rx-based PDSCH muting solution, where a UE receives PDSCH from a picocell when an interfering macrocell transmits an ABS. Instead of puncturing the REs suffering the CRS interference at the UE receiver, the transmitter at the pico eNodeB mutes the polluted REs during the PDSCH transmission, as shown in Figure 6.8.

The Tx-based PDSCH muting solution can be implemented in the resource mapping module and the rate matching module of the transmitter processing as shown in Figure 6.9. In the resource mapping module, the REs in the PDSCH region that suffer from strong CRS interference are muted and not used for carrying the PDSCH data. In the rate matching module, the PDSCH coding rate needs to be adjusted by taking into account those extra muted PDSCH REs.

From the signaling perspective, the UE needs to know the PDSCH muting pattern configured by the eNodeB so that it can perform resource element demapping accordingly. Note that the PDSCH muting pattern can be different in different subframes, as this depends on the CRS patterns and the MBSFN configurations of the interfering cells. The network has to convey this information to the UE. In the case of multiple dominant interferers, the network can potentially rate match around more than one of them (but not too many for reasons explained previously). The signaling should therefore support all these possibilities.

Consider a scenario where a picocell suffers strong interference from two neighboring macrocells that transmit their CRS in different locations. An example of the PDSCH muting patterns for the picocell to handle different CRS interference situations is shown in Figure 6.10. In this example, PDSCH muting pattern 1 corresponds to the case when there are no interfering cells. PDSCH muting patterns 2 and 3 can be used to mitigate the CRS interference from one macrocell with two different values of CRS shifts. PDSCH muting pattern 4 can be used to handle the CRS interference from both the macrocells. For example, if a subframe is configured as an MBSFN ABS by the two macrocells, PDSCH muting pattern 1 will be used in this subframe. In a subframe that is configured as a normal ABS by one macrocell and a MBSFN ABS by the other cell, PDSCH muting pattern 2 or 3 will be selected.

The Tx-based PDSCH muting solution can only handle the CRS interference in the non-colliding CRS case. In principle, if a set of REs can be punctured for the CRS interference

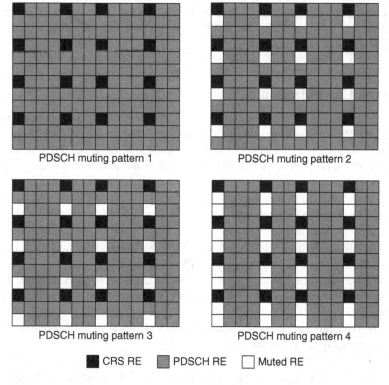

Figure 6.10   An example of PDSCH muting patterns configured by higher-layer signaling

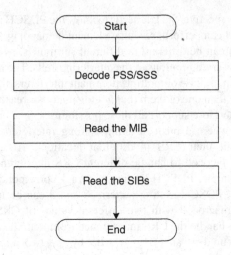

Figure 6.11   The cell detection procedure

mitigation in the Rx-based solution, the same set of REs can be muted in the Tx-based solution. Unlike the Rx-based solution, the Tx-based solution incurs no loss in information as the same amount of information is rate matched to yield a higher-rate transmission. Tx-based solutions are therefore generally expected to yield better performance. The higher-rate transmission also increases error rate and the final performance may depend on the specific deployment conditions, however.

6.4   Weak Cell Detection

When a UE is powered on, the first step is to detect its neighboring cells in order for association. After the RRC connection is established with the associated cell, periodic cell detection is performed for cell selection/reselection (for the UEs in the RRC_IDLE mode) and handover (for the UEs in the RRC_CONNECTED mode). In this section we show that inter-cell interference can severely affect these procedures; ICIC mechanisms are therefore needed.

To understand this first consider the procedure for cell detection which is shown in Figure 6.11. As discussed in Chapter 2, this can be summarized as follows.

Step 1: The UE detects the PSS/SSS and determines the exact carrier frequency, cell ID, radio frame boundary, and CP length after PSS/SSS detection.

Step 2: The UE reads the Master Information Block (MIB). The MIB contains the essential system information of the target cell, such as transmission bandwidth and number of antenna ports.

Step 3: The UE reads the System Information Blocks (SIBs). The SIBs contain other system information except for the MIB.

As the first step of cell detection, it is important for the UE to detect the PSS/SSS of the target cell. In a LTE/LTE-Advanced system, the PSS/SSS are transmitted in the same time-frequency locations in all cells. This means the interference affecting the PSS/SSS detection consists of the PSS/SSS signals from the neighboring interfering cells. If the target cell and its neighboring cells are aligned at the radio frame level and have the same CP length, they are also aligned in subframe level. In this case the PSS/SSS interference is severe and will degrade the performance. However, it is possible to perform the PSS/SSS interference cancellation before decoding the PSS/SSS from the target cell [8]. The flowchart of the PSS/SSS interference cancelation is shown in Figure 6.12, which is defined as follows.

Step 1: Decode the PSS/SSS from the strongest interfering cell.
Step 2: Estimate the interfering channel by assuming that the PSS/SSS transmitted from the interfering cell is known.
Step 3: Estimate and cancel the PSS/SSS interference from the strongest interfering cell.

The above procedure can be repeated to cancel the PSS/SSS interference from multiple interfering cells. Afterwards, the UE can perform normal PSS/SSS decoding for the target cell.

The same PSS/SSS interference cancelation scheme can also be used in the handover process. Consider the scenario where a macrocell hands over a UE to a picocell with a large CRE bias. The UE has to detect the PSS/SSS from the picocell in order to proceed to the handover process. However, the received PSS/SSS from the picocell is weak due to the PSS/SSS interference from the macrocell. In this case, the UE can cancel the PSS/SSS interference from the macrocell before detecting the PSS/SSS from the picocell.

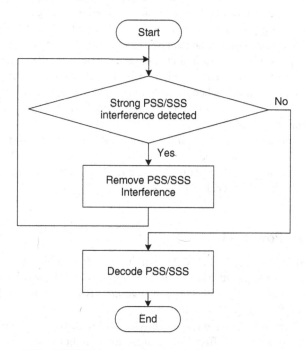

Figure 6.12   PSS/SSS interference cancellation for weak cell detection

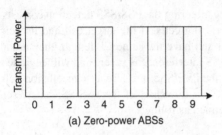 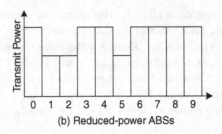

(a) Zero-power ABSs   (b) Reduced-power ABSs

Figure 6.13 Reduced-power ABS

The second step of cell detection is to read the MIB, which is carried on the PBCH. Similarly to the PSS/SSS transmission, a PBCH interference cancelation technique can be employed by the UE to detect the PBCH from a weak cell, which is summarized as follows.

Step 1: Detect the PBCH from the strongest interfering cell.
Step 2: Estimate the interfering channel by assuming that the PBCH transmitted from the interfering cell is known.
Step 3: Estimate and cancel the PBCH interference from the strongest interfering cell.

This procedure can be repeated to cancel the PBCH interference from multiple interfering cells.

There are certain restrictions on the network deployment to employ the PSS/SSS/PBCH interference cancelation solution: (1) all cells are required to be aligned in radio frame level; and (2) all cells are aligned in subframe level, which requires them to have the same CP length. These requirements ensure that the PSS/SSS signals are transmitted at exactly the same locations from all cells. If these requirements are not satisfied, the PSS/SSS interference cancelation is not applicable. In this case, the UE needs some assistance from the network side to detect weak cells. For example, when a macrocell hands over a UE to a picocell, it can provide the UE with the synchronization and system information of the picocell via higher-layer signaling, such that the UE does not need to decode the PSS/SSS/PBCH from the picocell.

The last step of cell detection is to read the SIBs, which are carried on the PDSCH. In order to decode the SIBs from a weak cell, it is necessary that the weak cell and the interfering cells coordinate utilization of PDSCH resources such that the interfering cells do not utilize resources used for SIBs in the weak cell.

6.5 Non-Zero-Power ABS

In LTE-Advanced Release 10, the PDSCH transmit power for a cell configuring ABS is strictly set to zero. In Release 11, this restriction has been removed. This means that a cell can transmit data to its associated UEs with low power in an ABS. We refer to this subframe as a non-zero-power ABS.

The difference between a non-zero-power ABS and a zero-power ABS is illustrated in Figure 6.13. In this example, subframes 1, 2, and 5 are configured as ABSs by a cell. Instead of not scheduling any UEs in the ABS, as in Release 10, the cell schedules some UEs and

Table 6.3 RE power reduction range

Modulation scheme	Maximum power reduction (dB)
QPSK	6
16QAM	3
64QAM	0

transmits to them at non-zero but reduced power in a Release 11 non-zero-power ABS. As a result, the cell has more resources for its own UEs and, at the same time, does not significantly cause interference to its neighboring cells because of the reduced PDSCH power. A higher ABS ratio is usually required for the non-zero-power ABS in order to keep the interference to the neighboring cells at the same level as the zero-power ABS.

Generally speaking, the non-zero-power ABS can achieve better system performance than the zero-power ABS [9, 10]. The reason for this is that it allows a cell more flexibility in adjusting its transmit power and configuring the non-zero-power ABSs. There are certain limitations related to practical implementation issues that affect the performance of non-zero-power ABS, however.

The most important factor affecting the performance of non-zero-power ABS is the power reduction range in ABSs. Ideally, the transmit power in ABSs can be any value below the maximum transmit power. However, due to the hardware limitations of the eNodeB, the maximum power reduction is limited and varies depending on the modulation schemes as shown in Table 6.3 [11].

Another important factor affecting the performance of the non-zero-power ABS is the knowledge of the exact PDSCH transmit power in the ABSs. The UE needs this information for the following two purposes [12].

- For the purpose of CSI measurement and feedback, the UE needs to know the PDSCH power; otherwise, the UE cannot compute the CQI accurately. The manner in which the UE is informed about the PDSCH transmit power depends on the transmission mode. In TMs 1–8, the UE derives CSI measurement and feedback based on CRS, which is transmitted with constant power in each subframe regardless of whether it is an ABS. The PDSCH transmit power to CRS transmit power ratio is signaled to the UE. Similarly, in TMs 9–10, the UE performs CSI measurement and feedback based on Channel State Information Reference Signal (CSI-RS), which is transmitted with constant power periodically. The relative power ratio of PDSCH and CSI-RS is also signaled to the UE.
- For the purposes of demodulation, the UE needs to know the PDSCH power. In TMs 1–6, the PDSCH transmit power is needed for QAM demodulation. In TMs 7–10, the PDSCH power information is not necessary for demodulation because DMRS is used, which is transmitted with the same power as PDSCH. Note that there is no need for power indication for QPSK modulation.

In Release 10, the ratio of the PDSCH to CRS/CSI-RS power for the non-ABS subframes is signaled to the UE via higher-layer signaling. A straightforward way to support Release 11 non-zero-power ABS is to additionally signal the power ratio of the PDSCH and CRS/CSI-RS for ABS subframes. The concept of restricted measurement sets can be directly applied to

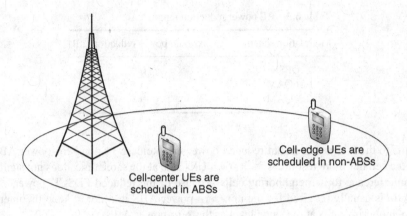

Figure 6.14   UE grouping for non-zero-power ABS

non-zero-power ABSs, that is, the UE measures and reports the CSI for ABSs and non-ABSs separately. The additional signalling can help the UE to perform restricted CSI measurement/feedback and demodulation in ABSs.

The non-zero-power ABS can also be implemented without introducing new signaling. For example, the UE applies the received power ratio of PDSCH and CRS for non-ABSs to ABSs for CSI feedback and measurement in the ABS subframes. This solution will cause CSI (especially CQI) feedback inaccuracy. To deal with CSI feedback inaccuracy, the eNodeB needs to adjust the UE-reported CSI based on the power reduction in ABSs. Alternatively, to avoid demodulation failure, the eNodeB needs to use only QPSK for PDSCH modulation in ABSs, even if the received CSI feedback suggests a higher-order modulation.

Another way to implement non-zero-power ABS without new signaling is to divide the UEs into two groups as shown in Figure 6.14: cell-center UEs and cell-edge UEs. The eNodeB schedules cell-center UEs and cell-edge UEs only in ABSs and non-ABSs, respectively. Furthermore, the power ratio of PDSCH and CRS/CSI-RS for ABSs is signaled to cell-center UEs, while the power ratio of PDSCH and CRS/CSI-RS for non-ABSs is signalled to cell-edge UEs.

The performance gain of the non-zero-power ABS over the zero-power ABS reduces with the above hardware and signal constraints. The following variations of non-zero-power ABSs are evaluated.

- Scheme 1 (without new signaling): QPSK only and 6 dB power reduction are used in ABSs. Only Release 10 signaling is assumed. Fixed Rank 1 transmission is used to simplify the eNodeB adjustments on the reported PMI.
- Scheme 2 (with new signaling): QPSK with 6 dB power reduction is used in ABSs. The power ratio of PDSCH to CRS for ABSs is signaled to UEs. The CSI measurement/feedback is performed based on the new signaling of power indication.
- Scheme 3 (with new signaling): QPSK/16QAM adaption with 3 dB power reduction is used in ABSs. The same new signalling as in Scheme 2 is used for CSI measurement/feedback.

The relative performance gain/loss of non-zero-power ABS over zero-power ABS is shown in Table 6.4 and Table 6.5 in terms of cell-average throughput and cell-edge throughput,

Table 6.4 Cell-average throughput comparison between non-zero-power ABS and zero-power ABS

CRE Bias (dB)	Release 10 Signalling	Scheme 1	Scheme 2
6	−5%	−2.8%	−3.2%
9	−7.2%	−4%	−3.88%

Table 6.5 Cell-edge throughput comparison between non-zero-power ABS and zero-power ABS

CRE Bias (dB)	Release 10 Signalling	Scheme 1	Scheme 2
6	13.3%	16.3%	27%
9	11.2%	13.1%	14.07%

respectively. It can be seen that there is performance loss in cell-average throughput although there is a throughput gain for the minority of the UEs in the cell-edge. Limited performance gain is therefore obtained by introducing new signaling of power indication. Based on the evaluation results, it was decided not to introduce any new signaling in Release 11 to support the non-zero-power ABS, which was left to proprietary implementation.

References

[1] NTT DoCoMo (2011) R1-113290, Investigation on Performance Improvement by CRE Considering CRS Interference. NTT DoCoMo, Technical Report, October 2011.

[2] Soret, B., Wang, Y., and Pedersen, K.I. (2012) CRS interference cancellation in heterogeneous networks for LTE-Advanced downlink. In *Proceedings of IEEE International Conference on Communications (ICC)*, pp. 6797–6801, June 2012.

[3] LG Electronics (2011) R1-113270, Performance Evaluation for FeICIC. LG Electronics, Technical Report, October 2011.

[4] Hitachi (2011) R1-113062, Performance Evaluation in Heterogeneous Networks Considering Crs Interference. Hitachi, Technical Report, October 2011.

[5] Texas Instruments (2011) R1-113244, Transmit-side Signalling Enhancements Targeting Het-Nets for Rel-11. Texas Instruments, Technical Report, October 2011.

[6] Nokia Siemens Networks and Nokia (2011) R1-113139, Considerations on Tx Based Enhancements for Rel-11. Nokia, Technical Report, October 2011.

[7] QualComm (2011) R1-113384, On Transmitter Based Solutions for eICIC. QualComm, Technical Report, October 2011.

[8] Damnjanovic, A., Montojo, J., Wei, Y., Ji, T., Luo, T., Vajapeyam, M., Yoo, T., Song, O., and Malladi, D. (2011) A survey on 3gpp heterogenous networks. *IEEE Wireless Communications Magazine*, **18**(3), 10–21.

[9] Huawei, HiSilicon (2011) R1-113635, Performance Evaluation of Feicic with Zero and Reduced Power ABS. Huawei, Technical Report, November 2011.

[10] Panasonic (2011) R1-113806, Performance Study on ABS with Reduced Macro Power. Panasonic, Technical Report, November 2011.

[11] 3GPP (2013) Evolved Universal Terrestrial Radio Access (E-UTRA); Base Station (BS) Radio Transmission and Reception 3GPP TS 36.104 v.11.5.0. Third Generation Partnership Project, Technical Report, July 2013.

[12] Huawei, HiSilicon (2012) R1-122522, Evaluation of Reduced Power ABS Based on RAN4 Feedback. Huawei, Technical Report, May 2012.

7

Release 11 Further Enhanced ICIC: Remaining Topics

This chapter discusses two important Release 11 eICIC mechanisms: interference coordination for data channels using multiple carriers and interference coordination for control channels by enhancing the structure of the legacy control channel, PDCCH.

The eICIC solutions in Release 10 discussed in Chapter 5 mainly focus on the co-channel interference scenario, where the macrocells and small cells are deployed in the same carrier. With the opening up of new frequency bands for LTE, future LTE eNodeBs will be capable of operating over multiple carriers. The expected popularity of such multi-carrier eNodeBs calls for the development of carrier-based ICIC solutions. In one such solution, the eNodeB can manage its interference by performing dynamic activation/deactivation of carriers based on the load and interference conditions in each carrier. The details of the ICIC techniques based on carrier activation/deactivation in the multi-carrier macro–pico deployment scenario is discussed in Section 7.1.

As the capacity of the data channels in LTE has improved radically due to enhancements in link and system level algorithms, the capacity of the legacy control channel PDCCH has become the 'bottleneck' in improving overall system performance. A new control channel, called Enhanced Physical Downlink Control Channel (EPDCCH), was therefore introduced in Release 11 to improve the control channel outage and capacity performances. We review the basic structure of EPDCCH and provide some examples of how to use it for inter-cell interference coordination in Section 7.2.

7.1 Carrier-Based Interference Coordination

An eNodeB can select a subset of carriers for offering LTE service from the set of available carriers that were assigned to it during the network planning process. These carriers are referred to as *operational carriers* and the process as *operation carrier selection* (OCS). In the multi-carrier macro–pico deployment scenario, the eNodeBs can utilize OCS processes to perform autonomous carrier activation/deactivation based on the interference environment and their traffic loads.

Heterogeneous Networks in LTE-Advanced, First Edition. Joydeep Acharya, Long Gao and Sudhanshu Gaur.
© 2014 John Wiley & Sons, Ltd. Published 2014 by John Wiley & Sons, Ltd.
Companion Wesite: www.ltehetnet.com.

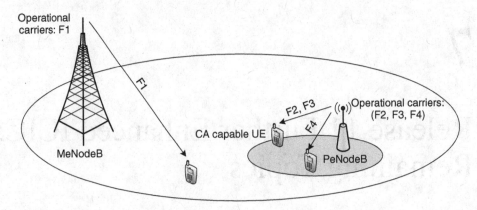

Figure 7.1 Illustration of CB-IC mechanisms, namely, operational carrier selection and UE-specific carrier selection in a multi-carrier macro–pico deployment scenario

Figure 7.1 shows an example deployment where potential operational carriers F1–F4 are assigned to the macro and pico eNodeBs during the network planning stage. The macro eNodeB (MeNodeB) and the pico eNodeB (PeNodeB) later select F1 and F2–F4 as their operational carriers, respectively, so that their transmissions do not interfere with each other.

In addition to OCS, an eNodeB can also perform UE-specific carrier selection [1] for the UEs without Carrier Aggregation (CA) capability by assigning a single carrier from the set of its operational carriers. For each CA-capable UE, the eNodeB selects one carrier as its Primary Cell (PCell) and one or more carriers as its Secondary Cells (SCells) as illustrated in Figure 7.1. The selection of the PCell/SCell(s) for a particular UE depends on the interference situation and traffic load on each carrier. Note that a non-CA-capable UE initially selects a carrier by itself to establish RRC connection. After the RRC connection is established, the eNodeB can hand over the UE to other carriers. Similarly, a CA-capable UE initially chooses its PCell by itself to establish RRC connection. After the RRC connection is established, the eNodeB can configure its SCell(s) via dedicated RRC signaling.

7.1.1 Operational Carrier Selection

The general procedure of carrier selection is shown in Figure 7.2. When a pico eNodeB (PeNodeB) is powered on, the Operation, Administration, and Maintenance (OAM) system configures operational carriers depending on the interference/load condition of the whole macro neighborhood. The OAM system can monitor the different levels of interference and load statistics as follows.

- Statistics from each eNodeB such as PRB utilization, transmit power, and handover parameter settings. The eNodeB reports these statistics for each operational carrier that is assigned to it.
- Statistics from each UE such as RSRP, RSRQ, CSI, handover measurement, and reconnection establishment statistics.

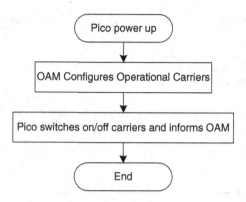

Figure 7.2   General procedure for carrier selection

Next the PeNodeB is responsible for activation/deactivation of the OAM-assigned operational carriers. The carriers could be activated/deactivated by the PeNodeB based on the following information.

- The carrier information of its neighboring eNodeBs, including carrier frequency (for uplink and downlink) and carrier bandwidth.
- The interference information of its neighboring eNodeBs such as Relative Narrowband Transmit Power (RNTP) and ABS pattern. RNTP is defined in Release 8 to support frequency domain ICIC. It contains 1 bit per RB, indicating whether the transmit power on the RB is greater than a pre-defined threshold. The PeNodeB can decide to deactivate a carrier suffering strong interference from its neighboring cells in the low- to medium-load situation.
- The load information of its neighboring eNodeBs, including hardware load, S1 Transport Network Layer (TNL) load, radio resource status, and ABS status. For example, a PeNodeB may decide to deactivate a carrier if it is underloaded and inform the neigboring PeNodeBs of this decision via X2. Alternatively, if the load at the eNodeB becomes very high, it may decide to activate the remaining carriers and inform the neighboring eNodeBs of the decision. This information exchange helps to achieve optimal use of the spectrum and can mitigate inter-cell interference.
- The report of RRM and CSI measurements from its associated UEs. In particular, a cell-edge UE can provide information about signal strength (i.e. RSRP) of its serving cell and neighboring cells, which is helpful for the eNodeB to understand the cell-edge interference levels.

Given the fact that the OAM system can monitor the whole macro neighborhood, it has a better understanding of the load and interference on a given carrier. However, the timescale of the carrier activation/deactivation via the OAM system is large. The eNodeB can also activate/deactivate a carrier by itself which will be much faster than the OAM-based carrier activation/deactivation. The eNodeB does not have access to the global interference and load conditions that OAM has, however, which leads to sub-optimal decisions. In most cases, the OAM-based solution is adopted in conjunction with the eNodeB-based solution, that is, the OAM system restricts the number of carriers configured for a given eNodeB, and the eNodeB

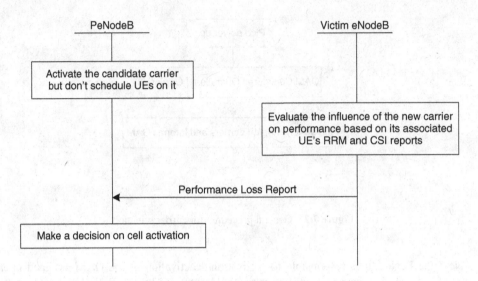

Figure 7.3 Carrier Activation Process

could further activate/deactivate carriers in the restricted range of carriers configured by the OAM system.

Carrier activation/deactivation performed by the eNodeBs needs enhanced frequency domain load and interference exchange between the eNodeBs involved via the X2 interface.

The carrier activation/deactivation decision can also be made based on the UE RRM and CSI measurement reports; the basic procedure is illustrated in Figure 7.3. Assume that initially one carrier is assigned to a PeNodeB by the OAM system and the load on this carrier becomes too high, subsequently necessitating the addition of a new carrier. If the PeNodeB adds a new carrier, it creates interference if the neighboring cells are also using that carrier to serve UEs. The picocell selects a new carrier and, at the same time, minimizes the interference caused to its neighboring cells as follows.

Step 1: The PeNodeB selects a candidate carrier to activate but does not schedule UEs on it. However, the pico-cell transmits CRS and other legacy controls signals.

Step 2: The UEs of the neighboring cell that are served on the candidate carrier evaluate the interference caused due to the PeNodeB via RRM and/or CSI measurement reports and report it to their serving cells.

Step 3: The eNodeBs in the neighboring cells evaluate the performance loss by comparing UE-reported measurements before and after the activation of the new candidate carrier.

Step 4: Each eNodeB communicates the performance loss due to activation of the candidate carrier over the X2 interface. Note that a new X2AP message is needed to allow this signaling.

Step 5: The concerned PeNodeB considers the impact of activating the candidate carrier by evaluating the trade-off between the reported performance loss of the neighboring eNodeBs and its own gain due to operation of the candidate carrier. Accordingly, it decides whether to make this candidate carrier operational.

While each eNodeB can make carrier switching decisions by itself based on the interference and load information from its neighboring cells, the carrier activation/deactivation can also be coordinated among eNodeBs via the backhaul. The following mechanism can be used for the coordinated carrier switching.

- An eNodeB informs its neighboring eNodeBs which carriers may be switched off to mitigate the interference.
- If an eNodeB suffers high interference in a carrier from a neighboring eNodeB, the eNodeB may inform the neighboring eNodeB about this by sending a carrier deactivation request.
- An eNodeB may inform its neighboring eNodeBs which carrier it will switch on by sending a carrier switch on request such that the neighboring eNodeBs can prepare for the sudden jump in interference levels. The neighboring eNodeB can respond with a request to delay the switching on if it estimates that doing so will cause high interference.

7.1.2   Primary and Secondary Cell Selection

Carrier Aggregation (CA) was introduced in Release 10 to allow a UE to connect to multiple component carriers simultaneously in order to increase the data rate. A CA-capable UE in the RRC_IDLE mode establishes an RRC connection with a serving cell, which automatically becomes its Primary Cell (PCell). As a result, different UEs may have different carriers as their PCells. After the RRC connection is established, the network can configure one or more Secondary Cells (SCells) for the UE to satisfy its traffic demand. Addition, removal, and reconfiguration of SCells can be performed depending on the traffic load and interference environment via dedicated RRC signaling. The network can only change the PCell of the UE via handover.

The network can configure different carriers as the PCell and SCells for a UE to manage the interference. For example, when the neighboring cells of an eNodeB mainly use carrier F1 as PCell, this eNodeB can select carrier F1 as an SCell for its associated UEs. This allows the possibility of quick deactivation of the SCell in case it starts to cause high interference at the neighboring PCell. To achieve such coordination, the following procedure is followed.

Step 1: An eNodeB receives the interference and PCell/SCell load information of the UEs of its neighboring eNodeBs via the X2 interface. The PCell/SCell load information can be implemented by extending the Resource Status Report of X2AP (this carries information about the number of UEs and PRB usage) to include information for multiple carriers and also information about whether the carrier is a PCell or a SCell.

Step 2: The eNodeB decides on the PCell/SCell configuration for a UE which guarantees PCell protection from inter-cell interference to allow reliable flow of control information. For example, if there are two carriers F1 and F2 such that F1 is protected from interfering neighboring cells whereas F2 is not, then the eNodeB can select F1 as the PCell and F2 as the SCell. It may also reduce the number of UEs using carrier F1 by deactivating the SCell on that carrier. Additionally, the load information of the PCell and the SCell can also help the eNodeB to decide on the assignment of PCell and SCells for a UE.

7.2   Enhanced PDCCH for Interference Coordination

In Release 10, the interference management for the Physical Downlink Control Channel (PDCCH) in the macro–pico co-channel deployment scenario was also addressed by time-domain coordination similar to interference management in PDSCH. The macrocells configured ABSs which also protected the PDCCH reception of the pico UEs. Given the fact that ABSs reduce the available radio resource and throughput of the macrocells [2], it is desirable to find a better way to coordinate the inter-cell interference for control signaling. Also, the legacy methods of choosing the first few OFDM symbols for PDCCH restrained the control channel capacity. Finally, control channel transmission in PDCCH did not benefit from the advanced transmission schemes used in PDSCH such as MIMO.

To overcome these issues, enhanced PDCCH (EPDCCH) was introduced in Release 11. EPDCCH is transmitted in the PDSCH region in units of PRB pairs.

In the following, we first provide a brief overview of the EPDCCH, mainly focusing on its resource allocation as specified in [3, 4, 5]. Afterwards, we illustrate several examples of using the EPDCCH to coordinate the inter-cell interference for control signaling.

One of the most important considerations for EPDCCH design is the coexistence with legacy UEs on the same carrier. In order to avoid affecting the legacy PDCCH, it was decided that the EPDCCH should be multiplexed with the PDSCH. This design allows for any unused EPDCCH resources to be assigned to legacy UEs for the PDSCH transmission. Figure 7.4 shows an example of the EPDCCH resource allocation in the PDSCH region. An EPDCCH can be scheduled in a pair of RBs in the same subframe. Having the EPDCCH and the PDSCH located in the same RB pair obviously introduces additional complexity, e.g. antenna port mapping, and is therefore not allowed. Rather, FDM of EPDCCH and PDSCH (the simplest approach) has been adopted as shown in Figure 7.4. This has a minimum impact on the PDSCH operation.

Having the EPDCCH transmission spanning an entire subframe reduces the time for the PDSCH decoding as the DCI decoding cannot be completed until complete reception of EPD-CCH. The PDCCH only occupies up to three or four OFDM symbols; a UE can therefore typically finish the PDCCH decoding and then start the subsequent PDSCH decoding before the end of the subframe. However, if the EPDCCH is used to carry the DCI, a UE has to wait

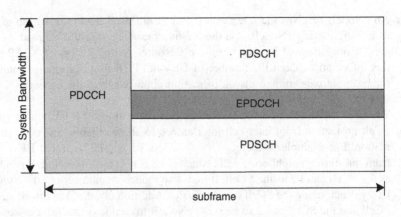

Figure 7.4   The concept of EPDCCH

several hundreds of microseconds after the end of the subframe to complete the EPDCCH decoding, which will give less time for the PDSCH decoding and HARQ ACK/NACK generation. The relaxation on the PDSCH decoding time was discussed in Release 11, but was ultimately left as an implementation issue.

The possible locations of the RB pair for the EPDCCH are UE-specific and configured via higher-layer signaling. If an RB pair is configured for the EPDCCH but not used for the EPD-CCH transmission, it can be used for the PDSCH transmission in order to improve resource utilization efficiency.

Recall that the basic resource unit for the PDCCH is Control Channel Element (CCE), which consists of 9 Resource Element Groups (REGs). The number of REs in each REG is fixed as 4; each CCE therefore has a total 36 REs. Each PDCCH is transmitted by one or more CCE. The number of CCEs used by a PDCCH is defined as aggregation level. In Release 10, the aggregation level can be chosen from 1, 2, 4, and 8, depending on channel condition.

Analogous to PDCCH, the basic resource unit for the EPDCCH is called the Enhanced CCE (ECCE), which consists of 4 Enhanced REGs (EREGs). The number of REs in each EREG varies depending on the presence of the legacy PDCCH and reference signals such as CRS and CSI-RS. In Figure 7.5, we illustrate the EREG to RE mapping in an RB pair for the case of normal CP length. It is shown that there are 16 EREGs in the RB pair. The EREG indices are sequentially mapped to REs, avoiding DMRS REs, in frequency and time domain. In Figure 7.5 a EREG is formed by the REs marked with the index of the EREG. Each EREG normally consists of 9 REs. Mapping the EREG indices in the way shown in Figure 7.5 ensures that the REs of each EREG are spread uniformly across the RB to maximize the time and frequency diversity. This ensures balanced performance for all EREGs.

The number of REs in each EREG can vary depending on the presence of the legacy PDCCH and other reference signals such as CRS and CSI-RS. In the example shown in Figure 7.5, if we assume that the PDCCH occupies 1 OFDM symbol and there are no CRS and CSI-RS, each of EGRGs 0–12 have 8 REs while EREGs 12–15 have 9 REs.

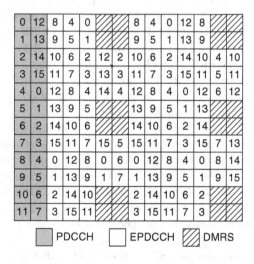

Figure 7.5 Enhanced resource element group

Table 7.1 Number of EREGs per ECCE

	Normal CP	Extended CP
Normal subframe	4	8
Special subframe configuration 1	8	8
Special subframe configuration 2	8	8
Special subframe configuration 3	4	8
Special subframe configuration 4	4	8
Special subframe configuration 5	8	8
Special subframe configuration 6	8	8
Special subframe configuration 7	8	8
Special subframe configuration 8	4	8
Special subframe configuration 9	8	8

In order to keep the performance of ECCEs similar and predictable, an ECCE is formed of either 4 EREGs or 8 EREGs, depending on subframe type as shown in Table 7.1. In principle, an ECCE has a larger number of EREGs in the subframe with a smaller number of available REs. The EREGs in a RB pair can be divided into the following 4 groups based on their indices:

EREG group 0: EREGs with indices 0, 4, 8, and 12.
EREG group 1: EREGs with indices 1, 5, 9, and 13.
EREG group 2: EREGs with indices 2, 6, 10, and 14.
EREG group 3: EREGs with indices 3, 7, 11, and 15.

If an ECCE consists of 4 EREGs, one of the above EREG groups will form the ECCE. If an ECCE consists of 8 EREGs, it will take EREG groups {0, 2} or {1, 3}. Such grouping rules ensure that all ECCEs in the same subframe have similar size and decoding performance.

The EREGs in an EREG group can come from the same or different RB pairs, depending on the EPDCCH configuration. If a localized EPDCCH is configured, it uses one or more ECCEs located within a single RB pair. If a distributed EPDCCH is configured, the EREGs in the EREG group(s) for one ECCE are located in different RB pairs. The example shown in Figure 7.6 shows how all EREGs of a localized EREG group come from the first RB pair, while the EREGs of a distributed EREG group comes from four different RB pairs. Note that the EREG groups within each RB pair are shown in a logical domain, which represents the same EREG to RE mapping as shown in Figure 7.6.

The localized EPDCCH transmission can improve the spectral efficiency when reliable CSI feedback is available. In Release 10, the eNodeB can configure its associated UEs to feedback CSI indicating sub-band channel condition for the purpose of frequency-selective scheduling and precoder selection. Since the EPDCCH uses a subset of RB pairs in the PDSCH region, the eNodeB can use the same CSI to schedule EPDCCH and uses the same optimal precoder for PDSCH. In the situation where reliable CSI feedback is not available, the distributed EPDCCH transmission can be configured to achieve robust performance by exploring frequency diversity.

We now provide several examples of how to use EPDCCH to coordinate the interference in control signaling. Consider a single-carrier macro–pico deployment scenario as shown in

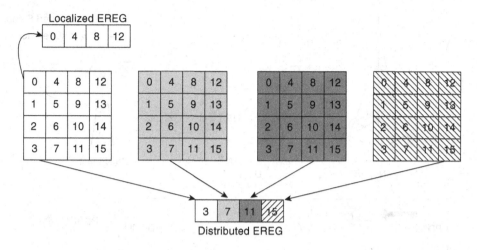

Figure 7.6   Localized and distributed EREGs

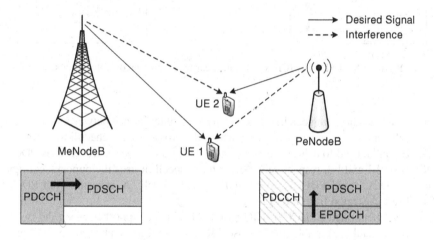

Figure 7.7   EPDCCH ICIC in the single-carrier macro–pico deployment scenario

Figure 7.7, where UE 2 associated to the picocell suffers strong interference from the macro-cell. To protect the control signal transmitted from the picocell, instead of the PDCCH the EPDCCH can be configured for this UE as shown in the figure. In the figure the linkage between PDSCH and the associated control channel (PDCCH or EPDCCH) is shown by an arrow. Accordingly, the macrocell does not schedule any of its UEs in the RB pair where the pico EPDCCH is being transmitted. As a result, the EPDCCH transmission is protected. Mean-while, having EPDCCH configured for the picocell reduces its interference to the macrocell on the PDCCH region (as less control information is being transmitted in the pico PDCCH), which will improve the PDCCH reception for macro UEs (e.g. UE 1).

Consider a multi-carrier macro–pico deployment scenario as shown in Figure 7.8, where the macrocell and the picocell operate on two carriers F1 and F2. In Chapter 5, we explained how

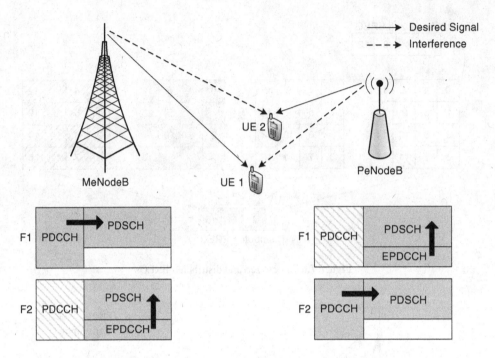

Figure 7.8 EPDCCH ICIC in the multi-carrier macro–pico deployment scenario

the cross-carrier scheduling can help to coordinate the interference between the macrocell and the picocell for control signaling. In case where the cross-carrier scheduling is not available, EPDCCH can be used to achieve the same goal as follows. The macrocell uses PDCCH and EPDCCH on F1 and F2, respectively. As for the picocell, it uses the opposite configuration, i.e. uses EPDCCH and PDCCH on F1 and F2, respectively. As a result, both PDCCH and EPDCCH are protected from interference.

The above-mentioned EPDCCH ICIC requires backhaul support between the two cells. If X2-based backhaul is available, the Release 8 RNTP signaling can be reused here for a cell to indicate the interference level it causes to its neighboring cells for each RB. After receiving the RNTP messages from its neighboring cells, the cell can identify the RBs which suffer less interference and configures them for EPDCCH. The location of EPDCCH can be configured via higher-layer signaling. If two cells are connected via near-ideal backhaul, for example fiber, a joint scheduler can be employed to coordinate the scheduling of PDSCH and EPDCCH for the two cells in each subframe.

In Release 12, new carriers with reduced control signaling called New Carrier Type (NCT) was introduced to improve the spectral efficiency [6]. For NCT, the PDSCH can start from the first OFDM symbol, which means the PDCCH is not mandatory [7]. If both the macrocell and picocell deploy NCT as shown in Figure 7.9, they can coordinate such that the EPDCCH are transmitted in protected RB pairs. This can completely remove the interference in the control channel.

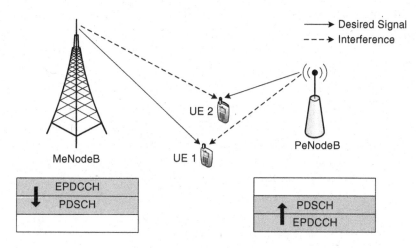

Figure 7.9 EPDCCH ICIC on New Carrier Type (NCT)

Note that EPDCCH can also be used to extend the capacity of the control channel in the macro–pico deployment scenario, where the macrocell and its neighboring picocells share the same cell ID (e.g. CoMP scenario 4, as explained in Chapter 8).

References

[1] 3GPP (2010) Evolved Universal Terrestrial Radio Access (E-UTRA); Carrier-Based HetNet ICIC Use Cases and Solutions 3GPP TR 03.024 v0.3.0. Third Generation Partnership Project, Technical Report, May 2010.

[2] Ye, S., Wong, S.H., and Worrall, C. (2013) Enhanced physical downlink control channel in LTE Advanced Release 11. *IEEE Communications Magazine*, **51**(2), 82–89.

[3] 3GPP (2012) Evolved Universal Terrestrial Radio Access (E-UTRA), Physical channels and modulation 3GPP TS 36.211 v11.0.0. Third Generation Partnership Project, Technical Report, September 2012.

[4] 3GPP (2012) Evolved Universal Terrestrial Radio Access (E-UTRA), Multiplexing and channel coding 3GPP TS 36.212 v11.0.0. Third Generation Partnership Project, Technical Report, September 2012.

[5] 3GPP (2012) Evolved Universal Terrestrial Radio Access (E-UTRA), Physical layer procedures 3GPP TS 36.213 v11.0.0. Third Generation Partnership Project, Technical Report, September 2012.

[6] Ericsson (2012) Rp-122028, New Carrier Type for LTE. Ericsson, Technical Report, December 2012.

[7] Huawei, HiSilicon (2013) R1-133817, Performance Evaluations of S-NCT vs. BCT. Huawei, Technical Report, August 2013.

Part Three

Coordinated Multi-Point Transmission Reception

8

Downlink CoMP: Signal Processing

8.1 Introduction

Coordinated transmission from multiple base stations to their UEs is very effective in mitigating the interference that occurs with the deployment of heterogeneous networks. From the early days of LTE Release 10, standardization activities in 3GPP have therefore attempted to standardize base station coordination. Several practical problems had to be solved before this could be made a reality. For example, theoretical algorithms for base station coordination [1] require ideal channel feedback at the transmitter, whereas in LTE channel feedback is quantized, delayed, and error-prone. Ideal base station coordination requires an ideal backhaul connecting the base stations for reliable communication between them, whereas practical backhauls have finite capacity and non-zero delay. Further, LTE Release 10 UE is only synchronized with its serving base station, which is a problem if it has to receive data from another base station with a different symbol and subframe timing.

Significant advances have been made over the past few years to standardize base station coordination. New genres of base station coordination algorithms have evolved that take limited feedback and non-ideal backhaul into account. All these various activities are categorized under the term Coordinated Multi-Point Transmission/Reception (CoMP) [2]. The first version of CoMP was included in LTE Release 11. Various CoMP schemes for downlink transmission can be grouped into the following categories.

- **Joint Processing (JP):** In this category of CoMP schemes, the UE data are available at multiple base stations. To enable this, the serving base station has to share the UE data with the other base stations via backhaul. Multiple base stations can simultaneously transmit to a UE. Alternatively, the UE may receive its data from a single base station, which changes dynamically based on the channel conditions. The former category is called *Joint Transmission* (JT) and the latter *Dynamic Point Selection* (DPS). JT CoMP offers more possibilities of cooperation than DPS, but but the associated signal processing and implementation complexity is higher for the latter.

Heterogeneous Networks in LTE-Advanced, First Edition. Joydeep Acharya, Long Gao and Sudhanshu Gaur.
© 2014 John Wiley & Sons, Ltd. Published 2014 by John Wiley & Sons, Ltd.
Companion Wesite: www.ltehetnet.com.

- **Coordinated Scheduling/Beamforming (CS/CB):** In this category of CoMP schemes, each UE is served by its serving base station as in Release 8. The scheduling and beamforming decisions at each base station are made jointly via coordination. The cooperating base stations do not share UE data, however. The absence of data transfer reduces backhaul load and implementation complexity.

Hybrid CoMP categories that combine these individual schemes are also possible, such as JT+CS/CB. In this chapter we explore the CoMP schemes in detail. We cover the details of signal processing and the changes that were made in the LTE standard to support CoMP.

At this juncture, we compare CoMP with Release 10/11 eICIC and FeICIC that were described in Chapters 5 and 6, respectively. ICIC techniques also required base station coordination. Cells communicated among themselves and the dominant interferers suspended transmission in ABS subframes. CoMP includes many other kinds of coordination strategies, however, and the timescales of coordination in CoMP are much faster. Release 10/11 eICIC/FeICIC algorithms are *semi-static* in nature. This means that the ABS patterns are changed via RRC signaling at a slower timescale than the CoMP coordination parameters that are signaled *dynamically* (on a subframe by subframe basis) to the UE using PDCCH or ePDCCH. As observed [3], CoMP achieves significant gains over eICIC. The challenge is to realize the theoretical gains of Release 11 CoMP under practical signaling and deployment constraints.

CoMP is possible in both downlink (base station coordination while transmission) and uplink (base station coordination while reception). Since the data rate requirements in the downlink are higher and more coordination strategies are applicable for downlink transmission, the majority of CoMP-related standardization activities in 3GPP are for the downlink. This chapter and the next will therefore focus only on downlink CoMP.

8.2 CoMP Scenarios in 3GPP

We start with the various CoMP scenarios that are defined in 3GPP. These scenarios were decided after discussions between different industry groups, chiefly operators, who jointly identified the most likely heterogeneous network scenarios where CoMP schemes are likely to be implemented. These scenarios were extensively evaluated with different CoMP categories to understand the gains of CoMP under practical deployment constraints. These scenarios are described in the following sections.

8.2.1 Homogeneous Networks with Intra-Site CoMP

This scenario, also known as *Scenario 1*, assumes coordination among the three co-located cells (sectors) that are controlled by a single eNodeB. As the cells are co-located there is no delay in transmitting the UE data or in sharing the scheduling information between the three cells. The gains of any CoMP strategy in this deployment are limited as the set of cells that cooperate to transmit data to a UE (the *CoMP Cooperating Set* as will be explained in Section 8.3.3) are pre-determined and fixed. Due to the time-varying nature of the wireless channel, maximum gains from CoMP are expected when the CoMP cooperating set for a UE is formed dynamically and aligned to the fading conditions experienced by the UE.

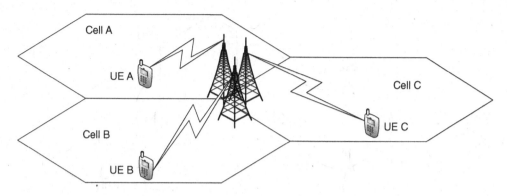

Figure 8.1   Homogeneous intra-site CoMP

Figure 8.1 depicts this scenario. Coordinated Scheduling/Beamforming (CS/CB) is an ideal CoMP strategy for this scenario. The eNodeB that controls the three cells jointly makes the scheduling and beamforming decisions in the three cells.

8.2.2   Homogeneous Networks with High-Power RRHs

Some operators prefer deployments where eNodeB functionalities are split into a baseband unit (BBU) which performs all baseband signal processing and a remote radio head (RRH) responsible for all RF operations. This scenario, also known as *Scenario 2*, represents DAS types of heterogeneous networks that were described in Section 1.2.1. The RRH has no intelligence of its own and merely transmits and receives radio signals provided to/from the BBU. Placing the RRHs in a distributed way maximizes the coverage of the eNodeB. Operators typically deploy optical fiber to connect the RRHs to the BBUs, thus ensuring a fast backhaul that is not capacity limited. This enhances CoMP performance.

An example of this scenario is shown in Figure 8.2. The figure shows an example of Joint Transmission (JT) CoMP transmission in this scenario, where multiple RRHs simultaneously transmit to a UE.

8.2.3   Heterogeneous Networks with Low-Power RRHs with Cell IDs Different from the Macro

This is an example of a heterogeneous network where numerous low-power nodes are overlaid over the macrocell coverage region. This scenario, also known as *Scenario 3*, describes a heterogeneous network with a mixture of public-access picocells and consumer-grade femtocells, which were described in Sections 1.2.2 and 1.2.3. A typical situation is depicted by Figure 8.3. Each of these nodes have their own cell IDs which means that they are independent eNodeBs. The presence of multiple eNodeBs in the macro coverage region increases network density. CoMP transmission from the eNodeBs can provide service to a large number of UEs. This scenario is therefore appropriate for networks that are likely to experience a high traffic load.

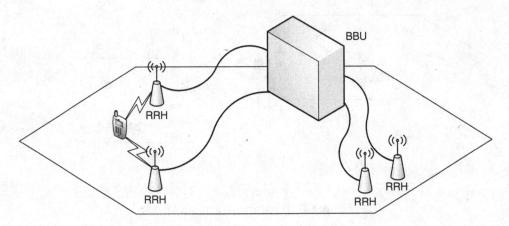

Figure 8.2   CoMP in Homogeneous Networks with high-power RRHs

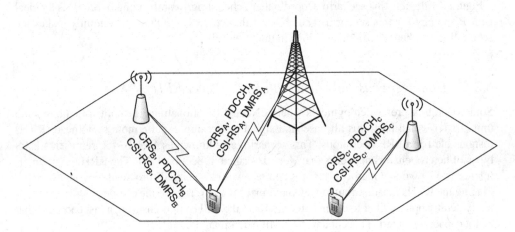

Figure 8.3   Heterogeneous networks with low-power RRHs with cell IDs different from the macro

Note that a large density of nodes leads to increased control channel interference, even when CoMP is applied. This could be more harmful than interference to the data channels and may limit the effectiveness of certain CoMP algorithms such as Joint Processing [4].

8.2.4   Heterogeneous Networks with Low-Power RRHs with Cell IDs the Same as the Macro

This scenario, also known as *Scenario 4*, is similar to that discussed in the previous section in terms of node deployment. However, all the nodes in the coverage region of the macro have the same cell ID. This means that all the nodes will transmit the same control signal (PDCCH) and common/broadcast reference signals (such as CRS). From the UE perspective, control channel interference is reduced. For the data channel reception, simultaneous transmission from multiple sources yields diversity gain at the UE. This is also referred to as a Single-Frequency Network (SFN) gain.

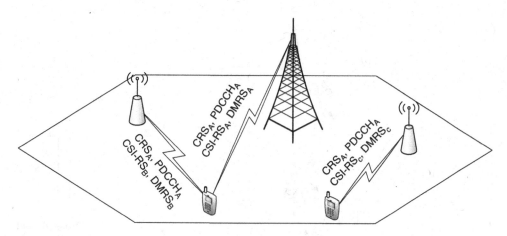

Figure 8.4 Heterogeneous networks with low-power RRHs with cell IDs the same as the macro

Since all nodes transmit the same PDCCH however, the control channel capacity of the network is reduced. A smaller number of UEs is consequently admitted into the network. This scenario is ideal for networks with a low or medium traffic load.

A typical scenario is depicted in Figure 8.4. Note that though the control and reference signals are common to all the nodes, UE-specific signals such as DMRS or CSI-RS can be different from different nodes. Different RRHs can therefore transmit data to different UEs simultaneously. This was not possible in Release 10 LTE, where DMRS and CSI-RS signal generations were also tied to the cell ID of the node. This restriction has however been removed in Release 11 and is discussed in more detail in the following chapter.

8.3 CoMP Sets

A *point* is defined as a set of geographically co-located transmit antennae [5]. The sectors of the same site correspond to different points. In the homogeneous network with high-power RRHs considered in Section 8.2.2, the RRHs correspond to different points even although there is only one eNodeB. Theoretically CoMP gains increase as the number of coordinating points increase. However, this imposes a huge burden on UE feedback overhead, backhaul traffic, and UE and base station computational complexity. A limited number of points should therefore be involved in CoMP coordination. The permissible number of points to ensure reasonable UE complexity for one CoMP-related operation such as multi base station feedback may be different from another operation. In LTE, different sets of points called *CoMP Sets* have therefore been defined for different CoMP-related processing.

The three main types of CoMP sets are depicted in Figure 8.5 and are described in the following sections.

8.3.1 *RRM Measurement Set/CoMP Resource Management Set*

The RRM measurement set consists of points for which the RRM measurements (e.g. RSRP and RSRQ) are performed. This was defined in Release 8 and hence no new procedure (either

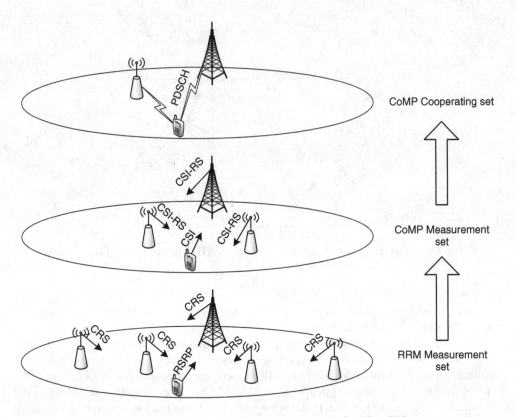

Figure 8.5   Illustration of different CoMP sets defined in 3GPP

in terms of indicating this set to the UE or extra UE feedback) is required to support CoMP. These measurements convey the long-term downlink channel quality from the various points. Based on the measurements, potential points can be selected for CoMP transmission to the UE. These selected UEs form the CoMP measurement set, described in the following section. The points in the RRM measurement set are RRC configured to the UE. A large number of points should be chosen in the RRM measurement set to prevent frequent RRC reconfigurations. A typical value is 32.

8.3.2   CoMP Measurement Set

Based on the RSRP measurements from the RRM measurement set, the network determines potential candidate points for CoMP transmission to the UE. This set of points is called the CoMP measurement set. Typically, these are points that have similar downlink gain to the UE. The UE measures and feeds back short-term channel state information (CQI and PMI/RI) for all points in the measurement set. This enables the base stations to decide the CoMP transmission parameters. A large measurement set increases UE computation and feedback complexity. The maximum size of the CoMP measurement set supported in Release 11 is 3.

Measurement set selection is based on RSRP which is measured from CRS. This poses a problem in CoMP Scenario 4 as all points transmit the same CRS, meaning that the UE cannot measure the link gains of the individual points. RSRP calculated from CSI-RS is a solution, but was not adopted in Release 11 as the sparse occurrence of CSI-RS leads to inaccurate RSRP estimates [6]. Measurement set formation is then either left to proprietary methods or based on uplink RS measurements. The inability to use CSI-RS RSRP will reduce CoMP gains in Release 11. However, Release 11 is the first step towards realizing CoMP and future releases are expected to include enhanced CoMP support.

8.3.3 CoMP Cooperating Set

The CoMP cooperating set consists of points that directly or indirectly participate in data transmission to the UE. The meaning of direct transmission of data is obvious, but points can also indirectly participate in transmission to a UE by aligning their own scheduling/beamforming decisions to minimize the interference to this UE. An example of this is Coordinated Scheduling/Beamforming (CS/CB).

8.4 CoMP Transmission in 3GPP

The three categories of CoMP transmission schemes, CS/CB, DPS, and JT, are discussed in detail in this section. Each CoMP category can be broadly distinguished based on the following criteria:

1. UE CSI feedback needed to support the CoMP category;
2. selection of UEs to be served via CoMP for a given group of cooperating base stations and/or CoMP cooperating set selection for a given UE; and
3. transmit precoder and MCS design.

These three criteria are interdependent and have to be jointly optimized for efficient CoMP operations. The details of this optimization can be understood by considering the specific CoMP categories, described in the following sections.

8.4.1 Coordinated Scheduling/Beamforming

In CS/CB, each UE is served by a single base station and the base stations coordinate their scheduling/beamforming decisions to mitigate the interference caused. To understand the details of CS/CB, let us consider the case where the coordinating base stations have been decided *a priori* and they have to schedule the UEs to be served via CoMP. As discussed in Section 8.2.1, such a CoMP strategy makes intuitive sense in homogeneous intra-site CoMP where the three co-located cells form natural candidates for a cooperating set for any UE which could be served via CoMP.

For purposes of simplicity, we consider a simple case of only two base stations: b_1 and b_2 transmitting to UEs u_1 and u_2 respectively. Let the channel between base station i and UE j be denoted $\mathbf{H}_{b_i u_j}$.

Let the base station i transmit to UE i using precoder $\mathbf{W}_i$. For simplicity, assume that only one layer of information is transmitted to UE i and hence $\mathbf{W}_i$ is a vector. Let the out of CoMP set interference and noise be denoted $\mathbf{I}_{\text{ext}}$ and $\mathbf{N}_0$, respectively. The achievable rate R_i of UE i is given by

$$R_1(u_1, u_2, \mathbf{W}_1, \mathbf{W}_2) = \log\left(\frac{\left|\mathbf{H}_{b_1 u_1} \mathbf{W}_1\right|^2}{|\mathbf{H}_{b_2 u_1} \mathbf{W}_2|^2 + \mathbf{N}_0 + \mathbf{I}_{\text{ext}}}\right) \tag{8.1}$$

$$R_2(u_1, u_2, \mathbf{W}_1, \mathbf{W}_2) = \log\left(\frac{\left|\mathbf{H}_{b_2 u_2} \mathbf{W}_2\right|^2}{|\mathbf{H}_{b_1 u_2} \mathbf{W}_1|^2 + \mathbf{N}_0 + \mathbf{I}_{\text{ext}}}\right). \tag{8.2}$$

CS/CB is given by the following joint optimization:

$$\max_{u_1 \in S_1, u_2 \in S_2, \mathbf{W}_1, \mathbf{W}_2} R_1(u_1, u_2, \mathbf{W}_1, \mathbf{W}_2) + R_2(u_1, u_2, \mathbf{W}_1, \mathbf{W}_2) \tag{8.3}$$

$$\text{s.t.}\quad Tr(\mathbf{W}_i \mathbf{W}_i^H) \leq P_i, \ i = 1, 2 \tag{8.4}$$

where S_i is the set of UEs associated with base station i that are eligible for CoMP. P_i is the transmit power constraint at base station i.

This is a non-convex optimization problem. It can be solved by heuristics such as separating the UE selection and precoder design problems [7].

Close-form solutions can however be obtained by considering metrics, other than sum rate, that also capture the essence of the CS/CB: *each point in the CoMP cooperating set is trying to simultaneously maximize the signal to its associated UE and minimize the co-channel interference to other UEs served by other points in the set*. In particular, it has been shown [8, 9] that signal to leakage noise ratio (SLNR) is a metric that adequately captures this essence. For the example provided above, the SLNR of the two base stations is

$$\text{SLNR}_1 = \frac{|\mathbf{H}_{b_1 u_1} \mathbf{W}_1|^2}{|\mathbf{H}_{b_1 u_2} \mathbf{W}_1|^2 + N_0} \tag{8.5}$$

$$\text{SLNR}_2 = \frac{|\mathbf{H}_{b_2 u_2} \mathbf{W}_2|^2}{|\mathbf{H}_{b_2 u_1} \mathbf{W}_2|^2 + N_0}. \tag{8.6}$$

In general, if there are K base stations in a system and base station i transmits to UE i using precoder $\mathbf{W}_i$, then the SLNR is defined

$$\text{SLNR} = \frac{|\mathbf{H}_{b_i u_i} \mathbf{W}_i|^2}{\sum_{k \neq i} |\mathbf{H}_{b_i u_k} \mathbf{W}_i|^2 + \mathbf{N}_0} \tag{8.7}$$

$$= \frac{|\mathbf{H}_{b_i u_i} \mathbf{W}_i|^2}{|\tilde{\mathbf{H}} \mathbf{W}_i|^2 + \mathbf{N}_0} \tag{8.8}$$

where $\tilde{\mathbf{H}} = [\mathbf{H}_{b_i u_1}^T, \cdots, \mathbf{H}_{b_i u_{i-1}}^T, \mathbf{H}_{b_i u_{i+1}}^T, \cdots, \mathbf{H}_{b_i u_K}^T]^T$. It has been shown [8] that the optimal choice of $\mathbf{W}_i$ that maximizes SLNR is given by

$$\mathbf{W}_i = \text{max eigenvector}\left(\left(N_0\mathbf{I} + \tilde{\mathbf{H}}^H\tilde{\mathbf{H}}\right)^{-1}\mathbf{H}_{b_i u_i}^H\mathbf{H}_{b_i u_i}\right). \tag{8.9}$$

In theory, if the cooperating base stations have complete information of all the downlink channels to various UEs, then they could jointly decide the precoders to enable CoMP. In practice this will be limited by the impartial channel knowledge at the transmitters because the UE channel state information feedback is quantized and suffers from delay.[1] The SLNR-based CS/CB processing therefore has to be modified to take into account quantized feedback [11–13].

In order to do this, first consider a generic UE feeding back channel state information to its associated base station. Let S be the set of subcarriers corresponding to a sub-band for which the UE feeds back PMI and CQI. The correlation matrix is defined

$$\mathbf{R} = \frac{1}{|S|}\sum_{s\in S}\mathbf{H}_s^H\mathbf{H}_s, \tag{8.10}$$

where $\mathbf{H}_s$ is the channel in subcarrier s. The UE PMI and RI are proportional to the eigenvector matrix of $\mathbf{R}$ and the CQI is proportional to the SINR, which in turn is proportional to the eigenvalue of $\mathbf{R}$. From the PMI/RI and CQI values from the UE, the base station can estimate the eigenvector and eigenvalues, $\hat{\mathbf{V}}$ and $\hat{\boldsymbol{\Sigma}}$, respectively. The base station can therefore reconstruct the approximate correlation matrix:

$$\hat{\mathbf{R}} = \hat{\mathbf{V}}\hat{\boldsymbol{\Sigma}}^2\hat{\mathbf{V}}. \tag{8.11}$$

For multiple base stations and UEs, each UE k can similarly feed back CQI and PMI for all base stations i in its CoMP measurement set, from which estimates of the downlink correlation matrix can be determined. This is called *per-point feedback* as the UE feeds back CSI without assuming any CoMP transmission. This is similar to Release 8 feedback but, unlike Release 8, the UE feeds back channel information for all base stations in the cooperating set.

The reconstructed correlation matrices are denoted $\hat{\mathbf{R}}_{b_i u_k}$; we focus our attention on base station i. The set of UEs who have base station i in their measurement sets are denoted $\mathcal{M}_i$. Base station i receives feedback from these UEs. The ideal precoder selection expression given by Equation (8.9) is modified for quantized feedback to yield

$$\mathbf{W}_i = \text{max eigenvector}\left(\left(N_0\mathbf{I} + \sum_{k\in\mathcal{M}_i, k\neq i}\hat{\mathbf{R}}_{b_i u_k}\right)^{-1}\hat{\mathbf{R}}_{b_i u_i}\right). \tag{8.12}$$

For MCS selection at the base station, the CQI reported by the UE cannot be used directly. This is because the UE-reported CQI is computed based on single-cell SU-MIMO transmission and does not take into account the out-of-cell interference due to the co-scheduled UEs in the other cells in the CoMP set. The base station can however estimate this interference based on feedback from all UEs [12]. Intuitively, the correlation between two PMIs that a UE reports for

---

[1] Note that for TDD LTE, channel reciprocity may be used in addition to UE feedback to determine the downlink channels. Since this is largely left to proprietary vendor implementation and outwith the scope of 3GPP standardization, we do not discuss it in this section. For issues related to channel reciprocity and CSI feedback in TDD see [10].

two different base stations indicates the correlation between the downlink channels of those base stations to the UE. The interference that one base station causes at the UE when the other base station is the serving cell scales with the value of correlation.

This intuition is utilized by the coordinating base stations to derive MCS based on CQI. Consider the scenario where each UE k feeds back a single (CQI, PMI) pair $(\gamma_{ik}, \hat{\mathbf{v}}_{ik})$ for base station i in its measurement set. Let base station i serve UE i. The base station has to choose a corresponding MCS $\tilde{\gamma}_{ii}$. The chosen MCS is defined

$$\tilde{\gamma}_{ii} = \frac{\gamma_{ii}}{1 + \sum_{k \in \mathcal{M}_i, k \neq i} \left| \hat{\mathbf{v}}_{ii}^H \hat{\mathbf{v}}_{ik} \right|^2 \gamma_{ik}}. \tag{8.13}$$

This shows that the MCS computation factors in the extra interference caused by the other co-scheduled UEs. In practice, for Equations (8.12) and (8.13), each base station i should choose only those UEs from the set $\mathcal{M}_i$ that are actually scheduled by the other cooperating base stations. This is usually not known *a priori*. The precoder and UE scheduling is therefore performed in a joint and iterative fashion (hence the nomenclature, coordinated scheduling, and beamforming). For each step of the iteration, base station i knows the UEs that were tentatively picked for scheduling by the other base stations in the last iteration step. The base station uses this information to pick its own UE and precoder/MCS values. All base stations perform these operations simultaneously. Usually the iterative procedure leads to good performance as shown by system level simulations [14].

8.4.2   Dynamic Point Selection

In this CoMP transmission category, only one point is dynamically selected to transmit data to the UE. As mentioned before, the UE data is available at multiple points to enable DPS. The transmitting base station can be potentially changed on a subframe-by-subframe basis. The intuition is that the association of a UE to its serving cell is based on the long-term channel characteristics but, due to the short-term channel fading, some other base station in the CoMP measurement set may have a higher downlink gain in a given subframe. Transmission from this base station would therefore increase the downlink rate in the given subframe. This would be more likely to happen for a cell-edge UE.

If DPS is implemented, the UE need not be aware that the transmitting base station has changed. As long as the new base station transmits with the same parameters as the serving base station (i.e. scrambled with the same UE RNTI, transmitted from the same antenna ports, and transmitting the required control information), the UE can decode the data without needing to know the identity of the transmitting base station. The CoMP operation is said to be *transparent* to the UE.

Transparency is a desirable feature in 3GPP as it implies that the receiver processing does not have to be changed. This reduces UE receiver complexity. However, this can also lead to loss of performance as the UE can only do better if it knows the cell ID of the transmitting base station and uses this information during the demodulation and decoding process. However, a well-designed DPS algorithm can ensure that this loss in performance is minimal [15].

The performance of DPS can be further enhanced by dynamically switching off the interfering base stations at certain sub-bands [16]. This is said to be DPS with DPB (dynamic point selection with dynamic point blanking). For example, a UE at the cell edge of a macro- and picocell has similar downlink channel strengths from both the cells. In a subframe where it is served by the picocell via DPS, the macrocell can mute its transmission to further enhance the performance. Both DPS and DPB can potentially be implemented at a sub-band level. The following example clarifies the various possibilities.

Example 8.4.1 *A typical deployment situation is shown in Figure 8.6. The picocell is the serving cell of UE A and the macrocell belongs to the measurement set of UE A. A possible corresponding sub-band level DPS/DPB pattern is shown in Table 8.1.*

It can be seen that the picocell transmits to UE A in the first four sub-bands. In the first of these two, the macro is muted. In the next two sub-bands however the macro transmits to UE B. This is because muting leads to wastage of a macro transmission opportunity and is inherently inefficient. Also note that in this example, DPS is frequency selective; UE A could be served by the macro in some other sub-band if the macro has a higher-link gain to it in that sub-band (e.g. sub-band 7).

With DPB the number of different *transmission hypotheses*, that is, combination of transmitting nodes and interferers, increases exponentially with the size of the measurement set. Even for the simple example given above, there are four transmission hypotheses as shown in Table 8.2. In the table, the term *External* denotes interference coming from base stations that are outwith the CoMP set.

Similar to CS/CB, in DPS/DPB the coordinating base stations need to know what MCS values can be decoded reliably at the UE for a given transmission hypotheses. This is based

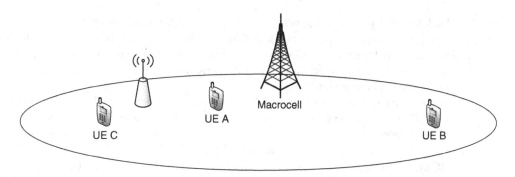

Figure 8.6 Example of a heterogeneous network to illustrate dynamic point selection with dynamic blanking

Table 8.1 Example of DPS/DPB. The various UEs scheduled by the macro and pico nodes in the different subbands are shown

	SB 1	SB 2	SB 3	SB 4	SB 5	SB 6	SB 7	SB 8	SB 9
Macro	Mute	Mute	UE B	UE B	UE B	UE A	UE A	UE B	UE B
Pico	UE A	UE A	UE A	UE A	UE C	UE C	Mute	UE C	UE C

Table 8.2 Different transmission hypothesis and corresponding configured CSI resources for a UE with two cells in its measurement set

	Signal	Interference	Example	Configured CSI resource
Hypothesis 1	Pico	External	Sub-band 1	CSI-RS: from pico, macro muted IMR: pico and macro muted.
Hypothesis 2	Pico	Macro + External	Sub-band 3	CSI-RS: from pico, macro muted IMR: pico muted, macro transmits
Hypothesis 3	Macro	External	Sub-band 7	CSI-RS: from macro, pico muted IMR: pico and macro muted
Hypothesis 4	Macro	Pico + External	Sub-band 6	CSI-RS: from macro, pico muted IMR: macro muted, pico transmits

on the CSI feedback from the UE. Recall that per-point feedback from the UE is feedback of single CSI report for each point. To illustrate this, consider Table 8.2. An example of per-point feedback from the pico- and the macrocell is feedback of CSI corresponding to hypothesis 1 for the pico and CSI for hypothesis 3 for the macro. Another example is feedback of CSI for hypothesis 2 for the pico and CSI for hypothesis 4 for the macro. However, per-point feedback will preclude feedback of CSI for all four hypotheses simultaneously. Per-point feedback from the UE for all base stations may not be sufficient as they do not capture the information of all possible transmission hypotheses that can arise during actual transmission. Other CSI reporting mechanisms were evaluated during the Release 11 CoMP standardization phase. These could be one or more of the following.

1. **Feedback for all hypotheses:** The network configures each UE with as many CSI resources as possible transmission hypotheses. Each CSI resource is a combination of a CSI-RS resource and IMR for channel and interference measurements, respectively. The UE can therefore measure the CSI values for all possible transmission hypotheses and feed them back to the serving base station.

   Interference measurement resources (IMR) were discussed in Chapter 4. Basically they are set of REs where a desired interference hypothesis can be measured. For example, IMR for UE A's hypothesis 2 is a set of REs where the picocell is muted and the macrocell transmits. The network decides the location of these REs in a frame and informs the UE via RRC signaling.

2. **Network estimation:** The network can configure a UE with a smaller number of CSI resources than the total number of possible transmission hypotheses. As an example, for Table 8.2 the UE is configured with CSI resources for hypotheses 1 and 3 and instructed to feedback their CSI. If the network needs CSI values corresponding to hypotheses 2 and 4, it estimates them via the CSI reports corresponding to hypotheses 1 and 3 [17].

3. **UE estimation:** The network can configure each UE with a smaller number of CSI resources than the total number of possible transmission hypotheses. As an example, for Table 8.2 the UE is configured with CSI resources for hypotheses 1 and 3. The network configures the UE to feedback CSI for all the hypotheses, however. Hence for hypotheses 2 and 4, the UE has to estimate the CSI without actual measurement. The UE emulates the interference for these transmission hypotheses and estimates the corresponding CSI [18].

The first approach leads to the most accurate results, but incurs a large signaling overhead. The base station has to configure more REs for transmitting CSI-RS and this reduces REs for PDSCH. There is more UE traffic in PUCCH/PUSCH for feeding back the CSI. It may also lead to redundancy as a particular transmission hypothesis (e.g. hypothesis 3) may never be used subsequently and thus there was no need to feed back the corresponding CSI. The other two approaches save on signaling overheads. They both try to estimate the CSI of a non-configured hypothesis from the CSI reports of other configured hypotheses. However, they are differentiated by which entity performs the estimation (the network or the UE). These approaches are illustrated in the following two examples.

Example 8.4.2 *The UE is configured with CSI resources for hypotheses 1 and 3 and is configured to report CSI only for these hypotheses. In this example, we show the method by which the network can estimate CSI for the other two hypotheses. We consider single-layer transmissions to the UE for ease of explanation. For single-layer transmission, the UE feeds back the following pairs of precoders and CQI values for hypothesis 1:*

$$\mathbf{W}_1 = \arg \max_{\mathbf{W} \in \mathcal{C}} \left| \mathbf{H}_p \mathbf{W} \right|^2 \tag{8.14}$$

$$CQI_1 = \frac{|\mathbf{H}_p \mathbf{W}_1|^2}{\mathbf{I}_{ext} + \mathbf{N}_0} \tag{8.15}$$

and the following for hypothesis 3:

$$\mathbf{W}_3 = \arg \max_{\mathbf{W} \in \mathcal{C}} |\mathbf{H}_m \mathbf{W}|^2 \tag{8.16}$$

$$CQI_3 = \frac{|\mathbf{H}_m \mathbf{W}_3|^2}{\mathbf{I}_{ext} + \mathbf{N}_0} \tag{8.17}$$

where $\mathbf{N}_0$ is the background noise power and $\mathbf{I}_{ext}$ is the external interference.[2]

The network has to estimate the CSI for hypotheses 2 and 4. First it can be seen that the transmitter in hypotheses 1 and 2 are the same; the same precoder could therefore be used for both. The interference is different, but choosing the same precoder has a negligible loss in performance. By a similar argument, the precoder for hypothesis 4 can be the same as that for hypothesis 3. We therefore have

$$\mathbf{W}_2 = \mathbf{W}_1, \ \mathbf{W}_4 = \mathbf{W}_3. \tag{8.18}$$

The question arises: how can the CQIs be estimated? In hypothesis 2, assume that the macro transmits to another UE with precoder $\mathbf{W}_{int}$ which causes interference at the UE. Let the correlation between the precoders $\mathbf{W}_3$ and $\mathbf{W}_{int}$ be ρ [19]. In this case,

$$CQI_2 = \frac{|\mathbf{H}_p \mathbf{W}_1|^2}{|\mathbf{H}_m \mathbf{W}_{int}|^2 + \mathbf{I}_{ext} + \mathbf{N}_0} = \frac{|\mathbf{H}_p \mathbf{W}_1|^2}{\rho^2 |\mathbf{H}_m \mathbf{W}_3|^2 + \mathbf{I}_{ext} + \mathbf{N}_0} = \frac{CQI_1}{\rho^2 CQI_3 + 1}. \tag{8.19}$$

The value of CQI_4 can be derived in a similar way.

---

[2] The quantity on the right-hand side of Equations (8.15) and (8.17) is actually a measure of SINR. The UE quantizes this information when it feeds back a CQI value. The base station will use the link adaptation mechanism to estimate the SINR value from reported CQI. These details are omitted from Equations (8.15) and (8.17) for the sake of simplicity. The overall effect of SINR quantization is an additional source of estimation error.

The value of correlation ρ affects the results. The CQI expression obtained here is essentially the same as that derived for CS/CB in Equation (8.13). This is because, in either case, the feedback was per point and the CQI to be estimated had to factor in the extra interference from nodes not present in the per-point CQI computations. In this case, the DPS scheduler at the base station can choose the co-scheduled UE (the one which is served by the macro) such that $\rho \sim 0$, i.e. it chooses a UE whose precoder is close to orthogonal.

By configuring CSI resources for hypotheses 1 and 3, it was easier for the network to estimate CSI for the other hypotheses. Hypotheses 1 and 3 involved CSI being computed without including interference from the points in the CoMP cooperating set. This form of CSI configuration has been extensively evaluated during the Release 11 phase and shown to yield good performance.

The norm-based method for selecting precoder and subsequent CQI computation are somewhat simplified approaches, optimal only for single-layer transmissions. For example, if the rank of the channel is greater than 1 then the UE will have to feed back a precoder for a higher number of layers and CQI for two codewords. The norm-based method will not extend naturally for this case. Instead, for each received layer, the UE has to compute the SINR that it can achieve with a specific receiver structure (such as a MMSE filter) and optimize the choice of precoders that would maximize the SINRs. The principles of CSI computation and estimation (either by network or UE) remain the same however, and the details are left as an exercise to the reader.

Example 8.4.3 *The UE is configured with CSI-RS resources for hypotheses 1 and 3. However, the UE is configured to report CSI for all hypotheses. In this example, we show how the UE can estimate the remaining CSI values. The values of $\mathbf{W}_1, \mathbf{W}_3, CQI_1, CQI_3$ are as given in the previous example. Further, as explained in the previous example, the UE can assume that $\mathbf{W}_2 = \mathbf{W}_1$ and $\mathbf{W}_4 = \mathbf{W}_3$. From the expression for CQI_2 given in Equation (8.19), the UE needs to know the macro interference term $\mathbf{I}_m = |\mathbf{H}_m \mathbf{W}_{int}|^2$. Note that the UE has already estimated $\mathbf{H}_m$ (from CSI-RS resource corresponding to hypothesis 3). It does not have any knowledge of the precoder $\mathbf{W}_{int}$. It can however make certain intelligent assumptions about that precoder and emulate $\mathbf{I}_m$. Examples of such assumptions include:*

- *the UE assumes that the network is likely to pick up a near-orthogonal precoder with high probability (this was also mentioned in the previous example as a potential network behavior); or*
- *the UE considers the set of all possible precoders and obtains an average value of $\mathbf{I}_m$.*

Once the UE has emulated the interference term $\mathbf{I}_m$, it can estimate the CQI as

$$CQI_2 = \frac{|\mathbf{H}_p \mathbf{W}_1|^2}{\mathbf{I}_m + \mathbf{I}_{ext} + \mathbf{N}_0}. \tag{8.20}$$

The value of CQI_4 can be determined in a similar way.

Let us compare this method for estimating CQI_2 to the network estimation method from Equation (8.19). The benefit of this approach is that the UE knows $\mathbf{H}_p$ which is not available at the network (which only knows $\mathbf{W}_1$). A potential problem with this approach is that there is an ambiguity about the UE behavior at the network side, meaning that the network does not know exactly how the UE had computed the CQIs for the additional hypotheses. If the network decides to alter the CQI values to derive the actual MCS (e.g. the network wants to implement MU-MIMO and wants to factor in the additional interference in the CQI computation), it becomes very difficult for it to do so. This could have been remedied by reporting the UE behavior or fixing it to a known value. This was deemed unreliable by 3GPP, however. Currently, UE emulation of interference and estimation of CSI for non-configured resources has not been standardized in Release 11. The network configures the UE with certain CSI resources and estimates the rest of the CSI values as and when needed.

8.4.3 Joint Transmission

Joint transmission is the most general form of CoMP when multiple base stations transmit data simultaneously to a UE. There is no unique way of implementing JT. Some common algorithms are discussed in this section. Although JT potentially offers the largest gains, UE feedback and base station processing complexity can increase. Practical constraints such as finite bandwidth and non-zero latency backhauls will affect the performance of JT more than other CoMP schemes. 3GPP implementation of JT assumes that the JT operation is transparent to the UE, as for DPS as discussed in Section 8.4.2. Before going into the details of the various possible JT algorithms, we first establish some definitions.

Definition 8.4.4 *Per-CSI-RS-resource feedback: In per-CSI-RS-resource feedback, the UE measures the channel and interference from the associated CSI-RS resource and IMR and reports back PMI/RI and CQI.*

Per-CSI-RS resource feedback introduces the possibility of a CSI-RS resource aggregated over multiple points. For example, consider a four-port CSI-RS resource that has been aggregated to measure the channels of two base stations each having two antenna ports. Another feedback type, namely per-point feedback, was previously considered for DPS/DPB. It can be re-defined as follows.

Definition 8.4.5 *Per-point feedback: Per-point feedback is a special case of per-CSI-resource feedback, where the associated CSI-RS resource is transmitted from a single point.*

Fundamentally, all JT schemes can be classified into two types [20]. For a macro–pico network, these are defined:

$$y = H_m W_m s + H_p W_p s + z \qquad \text{(streamperpoint)} \qquad (8.21)$$

$$y = H_m W_m s_1 + H_p W_p s_2 + z \qquad \text{(differentstreamsperpoint)} \qquad (8.22)$$

Although very similar from a theoretical point of view, there are important distinctions between the two based on practical constraints. Following a similar approach to DPS/DPB, we now consider the three methods for CSI reporting and estimation and examine how the two types of JT schemes are affected.

1. **Configured aggregated feedback:** In this case, the UE has the ability to directly measure the composite channel of the macro and pico, that is, $H_{JT} = [H_m, H_p]$. This is possible by configuring a CSI resource from the macro and pico to measure the aggregated channel.

For example, if both base stations have two antennae each, it is possible to configure a four-port CSI-RS resource such that the CSI-RS signal corresponding to the first two ports are transmitted by the macro and for the remaining two ports from the pico. This operation is transparent to the UE.

For this case, the distinction between the two types becomes redundant. The UE feeds back a precoder $\mathbf{W}_{JT}$ for the channel $\mathbf{H}_{JT}$. The JT transmission process can be modeled as

$$\mathbf{y} = \mathbf{H}_{JT}\mathbf{W}_{JT}\mathbf{s} + \mathbf{z}. \tag{8.23}$$

The individual precoders to be used at the macro and pico base stations, $\mathbf{W}_m$ and $\mathbf{W}_p$, can be derived from $\mathbf{W}_{JT}$ by expressing it as $\mathbf{W}_{JT} = [\mathbf{W}_m^T, \mathbf{W}_p^T]^T$. The CQI (therefore MCS) for single-layer transmission can be derived:

$$CQI_{JT} = \frac{|\mathbf{H}_{JT}\mathbf{W}_{JT}|^2}{\mathbf{I}_{ext} + \mathbf{N}_0}. \tag{8.24}$$

Note that configuration of this aggregated CSI resource is in addition to the other CSI resources that have to be configured for per-point feedback (such as the four mentioned in Table 8.2). This is because, during the feedback phase, the exact CoMP transmission strategy has not yet been decided. Cconfiguration of this extra CSI resource for capturing aggregated channel parameters therefore increases the overhead.

Can the JT CQI be estimated from per-point feedbacks? The following example discusses this issue.

Example 8.4.6 *Let the UE be configured with per-point hypotheses 1 and 3 as defined in Table 8.2. It feeds back per-point precoders $\mathbf{W}_m$ and $\mathbf{W}_p$. Consider that JT transmits the same stream from the two base stations. The received signal at the UE is*

$$\mathbf{y} = (\mathbf{H}_m\mathbf{W}_m + \mathbf{H}_p\mathbf{W}_p)\mathbf{s} + \mathbf{z}. \tag{8.25}$$

This is clearly sub-optimal however as the per-point channels may be such that the effective channels $\mathbf{H}_m\mathbf{W}_m$ and $\mathbf{H}_p\mathbf{W}_p$ do not add up in phase or, worse, are in opposite spatial directions and cancel the received energy. Consider that the Singular Value Decompositions (SVD) of the three channels are given by $\mathbf{H}_m = \mathbf{U}_m\boldsymbol{\Sigma}_m\mathbf{V}_m^H$, $\mathbf{H}_p = \mathbf{U}_p\boldsymbol{\Sigma}_p\mathbf{V}_p^H$ and $\mathbf{H}_{JT} = \mathbf{U}_{JT}\boldsymbol{\Sigma}_{JT}\mathbf{V}_{JT}^H$. These channel subspaces are related by $\mathbf{U}_{JT}\boldsymbol{\Sigma}_{JT}\mathbf{V}_{JT}^H = [\mathbf{U}_m\boldsymbol{\Sigma}_m\mathbf{V}_m^H, \mathbf{U}_p\boldsymbol{\Sigma}_p\mathbf{V}_p^H]$. From this expression, there is no simple relationship between $\mathbf{V}_{JT}$ and $(\mathbf{V}_m, \mathbf{V}_p)$, the subspaces on which the precoders are based.

The above example demonstrates the sub-optimality of per-point feedback. 3GPP studies during Release 11 phase considered other feedback schemes that estimate the JT CQI better while still keeping the feedback overhead less than that required for aggregated CSI feedback. Specifically, 3GPP had considered the possibility of the UE feeding back inter-point phase ϕ feedback [21] to enhance per-point CSI feedback. For inter-point phase feedback, the JT transmission equation (Equation (8.25)) is modified to

$$\mathbf{y} = (\mathbf{H}_m\mathbf{W}_m + \phi\mathbf{H}_p\mathbf{W}_p)\mathbf{s} + \mathbf{z}. \tag{8.26}$$

The inter-point phase carries quantized information to denote the correlation between the effective channel subspaces and is chosen by the UE so that $\mathbf{H}_m\mathbf{W}_m$ and $\phi\mathbf{H}_p\mathbf{W}_p$ are aligned spatially. The UE thus feeds back ϕ for each pair of configured CSI resources. If N bits are used for feedback and K is the size of the CoMP measurement set, then the total feedback budget would be equal to $(K-1)N$ bits. This is obtained by assuming that the phase of the channel subspace of one point in the set is the reference for all other points.

More generally, the inter-point phase information can be chosen in order to make the effective precoder $\mathbf{W}_{\text{InterPoint}} = [\mathbf{W}_m^T, \phi\mathbf{W}_p^T]^T$ as close to the aggregated JT precoder as possible. Due to shortage of time during the Release 11 evaluation process, no consensus could be reached about the quantitative gains of inter-point phase information. This has not been included in Release 11 CoMP but may be considered in the future.

2. **Network estimation:** Suppose that the UE has been configured to report per-point feedback of hypotheses 1 and 3 as given in Table 8.2. The network can choose the same per-point precoders for JT transmission. In the absence of inter-point phase information, the JT scheduler can perform UE selection based on how correlated the individual per-point precoders are. Depending on other MIMO processing such as MU-MIMO, the network can decide to alter the precoders from what is reported by the UE. Let the correlation between the reported and chosen precoders be given by ρ. When both base stations transmit the same data stream, the network derives the JT CQI [22, 23], which is used for the derivation of the MCS as follows:

$$\text{CQI}_{\text{JT}} = (\sqrt{\text{CQI}_1} + \rho\sqrt{\text{CQI}_2})^2. \tag{8.27}$$

However if the two base stations transmit different data streams, the achievable JT CQIs are

$$\text{CQI}_{\text{JT},1} = \frac{\text{CQI}_1}{\rho^2\text{CQI}_2 + 1} \tag{8.28}$$

$$\text{CQI}_{\text{JT},2} = \frac{\text{CQI}_2}{\rho^2\text{CQI}_1 + 1}. \tag{8.29}$$

3. **UE estimation:** The UE can also estimate the aggregated CQI values by evaluating Equations (8.27)–(8.29) since it knows the downlink channels. The drawback of this method is that it does not know the final precoders that will be picked up by the network. If it makes assumptions about the possible choice of precoders, it may be difficult to convey this information to the network without additional signaling. Hence, as with DPS, UE estimation methods have not been standardized for JT CoMP.

8.4.3.1 Multi-User JT

Dynamic SU/MU-MIMO switching has been identified as one of the key components to improve the downlink throughput performance. Multi-user JT [24] is a natural extension of MU-MIMO. The total number of transmit antennae increases when base stations collaborate, leading to the possibility of transmitting multiple layers of information, possibly to different UEs. MU-JT is similar to MU-MIMO in terms of UE feedback and base station precoder selections. Each UE can feed back per-point precoders and the network can perform UE selection by choosing UEs with orthogonal PMIs. Based on the actual precoders that the base station selects for the UE and for all other UEs that are co-scheduled with it, the network can adjust the CQI values.

Like MU-MIMO, performance of MU-JT can be improved by enhanced UE feedback such as channel direction feedback, where the UE feeds back information about the eigenvalues and eigenvectors of the downlink channel and companion feedback where the UE feeds back the precoder of the co-scheduled UE which would cause minimum interference to itself [25]. These enhanced feedback mechanisms have not been standardized in Release 11, but may be considered for future CoMP releases.

8.4.3.2 Scheduling for JT CoMP

The performance of JT heavily depends on the scheduler that decides which UEs will be served by the JT. It should be realized that, even if a UE is capable of being served by CoMP (i.e. it has multiple points in its CoMP cooperating set), it is possible that non-CoMP transmission from its serving cell improves the overall performance. Given that there are multiple sub-bands over which a UE or group of UEs may be scheduled via SU/MU-MIMO or CoMP, and a common MCS has to be chosen over all the sub-bands that a UE has been allocated, there is no optimal and elegant way to perform wideband CoMP scheduling.

There are many possible sub-optimal JT schedulers [26, 27, 28, 29, 30], based on intelligent heuristics that focus on identifying scenarios where JT would perform better than SU-MIMO. A possible JT scheduler [26] is reproduced below.

1. Determine the potential JT UEs and non-JT UEs based on the UE feedback.
2. For each point k, select the UE u_k with the best proportional fair metric $PF(k, u_k) = r(k, u_k)/R_{avg}(u_k)$ where $r(k, u_k)$ is the reported CQI when point k transmits to UE u_k and $R_{avg}(u_k)$ is the average rate of the UE.
3. Sort all the JT UEs u by their corresponding PF metrics $PF(i, j, u) = r(i, j, u)/R_{avg}(u)$, where $r(i, j, u)$ is the CQI achieved by the UE when it is served via JT by points i and j.
4. For each sub-band, sort the JT UEs. This could be performed based on the PF metrics.
5. For each JT-UE u in the determined sorting order and associated JT point pair (i, j), compare $PF(i, j, u)$ and $PF(i, u_i) + PF(j, u_j)$. Skip the UE u if either cell i or cell j have already assigned a UE for transmission in that sub-band. Note that since the set of JT UEs is a subset of all the UEs considered, either UE u_i or u_j could also be UE u.
6. If $PF(i, j, u) \geq PF(i, u_i) + PF(j, u_j)$ then assign JT transmission to UE u by points i and j in the given sub-band. Otherwise, assign SU-MIMO transmission for point i to UE i and point j to UE j.

This algorithm provides a fair balance between SU-MIMO and JT transmissions. Further optimizations are possible in step (4), or the PF metric could be replaced by some other metric such as instantaneous rates. Ultimately, a JT scheduler has to be extensively tested to see if it yields satisfactory performance for a given deployment scenario.

8.5 Comparison of Different CoMP Categories

Dynamic coordination in CoMP yields higher gains than the semi-static coordination of eICIC. To determine the exact quantitative nature of the gains, extensive evaluations have been performed by 3GPP. These results are captured in [5], a summary of which is provided in this section.

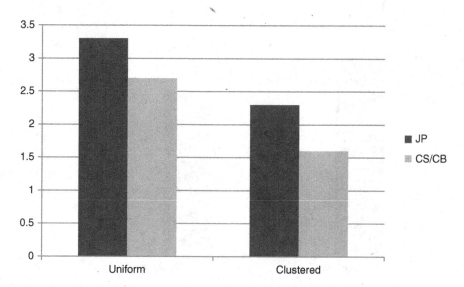

Figure 8.7 Percentage cell average performance gain of downlink CoMP in scenarios 3 and 4 with full buffer traffic model over Release 10 eICIC

The evaluations considered a multicell network with 57 macrocells and 4 small cells per macro area. Two UE scenarios were modeled. In the first scenario, UEs were uniformly distributed over the macro area. In the second scenario, about two-thirds of the UEs were clustered around the small cells that are assumed to be data hotspots. System level simulations were performed to compute the throughput for JP, CS/CB and Release 10 eICIC for the two scenarios.[3]

Figure 8.7 shows the average gains of all the UEs in the system. As expected, JP shows the best performance compared to CS/CB and eICIC. Among the three schemes, only in JP can an UE dynamically switch the serving base station(s). The probability that a UE is served by the base station(s) with which it has the maximum downlink signal strength is therefore maximized in JP.

More insight into the gains of CoMP can be understood by looking at those UEs whose throughput constitute the lowest 5% in the range of achievable throughputs for all UEs in the system. If the UEs are uniformly distributed over the entire macro area, the majority of the lowest 5% UEs will be located near the cell edge. Figure 8.8 shows that the JP and CS/CB CoMP offer about 52% and 20% average gains respectively over Release 10 eICIC for this scenario. However, when the majority of the UEs are clustered around the small cells, the gains of both JP and CS/CB CoMP are reduced.

---

[3] As should be clear by now, there is no unique JP, CS/CB, and eICIC algorithm specified in the 3GPP standards. Several important components such as scheduling and link adaptation are left to proprietary implementation, and these affect the performance of a CoMP or ICIC algorithm. The results presented here are averaged over evaluation results from different companies. The averaging process may not correctly capture the information in the different results. However, it is expected to correctly capture the relative gains of a class of CoMP over ICIC.

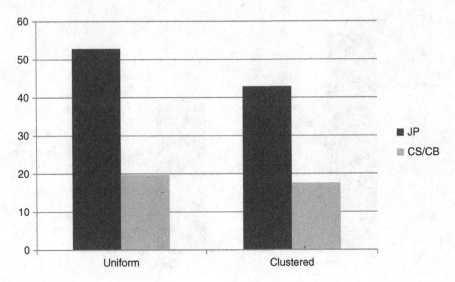

Figure 8.8 Lowest 5% user performance gains of downlink CoMP in scenarios 3 and 4 with full buffer traffic model compared to Release 10 eICIC

The underlying phenomenon which explains this is that there are less cell-edge UEs in the lowest 5% when more UEs are clustered around the small cells. The probability that some of the lowest 5% UEs are far away from the cell edge is higher for the clustered scenario. The following conclusions can therefore be drawn.

1. CoMP is most beneficial for cell-edge UEs. This is supported by the fact that CoMP gains (both JP and CS/CB) are in the range of 20% and higher if only cell-edge UEs are concerned, but fall to 5% and below when all UEs are taken into account in the computation of the cell average. However, these other UEs (not near the cell edge) experience much less inter-cell interference and achieve a sufficient throughput even without CoMP. CoMP therefore benefits those UEs which are in most need of help.
2. The gain of JP CoMP is more sensitive to the UE location. As seen from Figure 8.8, JP gain drops by 10% from uniform to clustered while that of CS/CB drops by about 2%.

References

[1] Karakayali, M.K., Foschini, G.J., and Valenzuela, R.A. (2006) Network coordination for spectrally efficient communications in cellular systems. *IEEE Wireless Communications Magazine*, **13**(4), 56–61.
[2] Sun, S., Gao, Q., Peng, Y., Wang, Y., and Song, L. (2013) Interference management through CoMP in 3GPP LTE-advanced networks. *IEEE Wireless Communications Magazine*, **20**(1), 59–66.
[3] Li, Y., Li, J., Li, W., Xue, Y., and Wu, H. (2012) CoMP and interference coordination in heterogeneous network for LTE-advanced. In *Proceedings of IEEE GLOBECOMM*, December 2012.
[4] Lee, D., Seo, H., Clerckx, B., Hardouin, E., Mazzarese, D., Nagata, S., and Sayana, K. (2012) Coordinated multipoint transmission and reception in LTE-advanced: Deployment scenarios and operational challenges. *IEEE Communications Magazine*, **50**(2), 148–155.

[5] 3GPP (2011) Coordinated multi-point operation for LTE physical layer aspects. 3GPP TR 36.819, v11.1.0. Third Generation Partnership Project, Technical Report.

[6] Qualcomm Incorporated, NTT DoCoMo (2012) R1-123688, on CSI-RS based RSRP feasibility in Rel-11. QualComm, Technical Report, August 2012.

[7] Jang, U., Son, H., Park, J. and Lee, S. (2011) CoMP CSB for ICI nulling with user selection. *IEEE Transactions on Wireless Communications*, **10**(9), 2982–2993.

[8] Sadek, M., Tarighat, A., and Sayed, A.H. (2007) A leakage-based precoding scheme for downlink multi-user MIMO channels. *IEEE Transactions on Wireless Communications*, **6**(5), 1711–1721.

[9] Qiang, L., Yang, Y., Shu, F., and Gang, W. (2010) SLNR precoding based on QBC with limited feedback in downlink CoMP system. In *Proceedings of Wireless Communications and Signal Processing (WCSP)*, October 2010.

[10] CATT (2011) R1-112956, CQI feedback for CoMP. CATT, Technical Report, October 2011.

[11] Motorola (2009) R1-093963, SCF-based COMP: Iterative scheduler algorithm and performance gain over single-point SU/MU beamforming. Motorola, Technical Report, October 2009.

[12] DoCoMo (2011) R1-110866, system performance for CS/CB-CoMP in homogeneous networks with high tx power RRHs. DoCoMo, Technical Report, February 2011.

[13] Intel (2011) R1-110252, coordinated beamforming schemes for CoMP. Intel, Technical Report, January 2011.

[14] Hitachi (2011) R1-111283, Preliminary Evaluation of DL Homogeneous CoMP. Hitachi, Technical Report, May 2011.

[15] Tan, Z., Zhou, W., Chen, W., Chen, S., and Xu, Y. (2011) A dynamic cell selection scheme based on multi-object for CoMP DL in LTE-A. In *Proceedings of International Conference on Communication Technology and Application (ICCTA)*, October 2011.

[16] Feng, M., She, X., Chen, L., and Kishiyama, Y. (2010) Enhanced dynamic cell selection with muting scheme for DL CoMP in LTE-A. In *Proceedings of IEEE Vehicular Technology Conference (VTC)*, May 2010.

[17] Hitachi (2012) R1-122699, CQI estimation for CoMP. Hitachi, Technical Report, May 2012.

[18] Ericsson (2012) R1-122837, CQI definition of UE emulated intra CoMP cluster interference. Ericsson, Technical Report, May 2012.

[19] LG Electronics (2011) R1-113276, CQI calculation for CoMP. LG Electronics, Technical Report, October 2011.

[20] NTT DoCoMo (2011) R1-113292, CSI feedback scheme for JT CoMP. NTT DoCoMo, Technical Report, October 2011.

[21] Huawei CATT, CMCC, HiSilicon, Hitachi, Intel, LG Electronics, New Postcom, Sharp, and ZTE (2012) R1-120866, Way forward on CSI feedback for CoMP. Technical Report, February 2012.

[22] Intel (2011) R1-112913, discussion on CQI definition in CoMP systems. Intel, Technical Report, October 2011.

[23] ZTE (2011) R1-113009, CQI computation for CoMP. ZTE, Technical Report, October 2011.

[24] Matsuo, D., Rezagah, R., Tran, G.K., Sakaguchi, K., Araki, K., Kaneko, S., Miyazaki, N., Konishi, S., and Kishi, Y. (2012) Shared remote radio head architecture to realize semi-dynamic clustering in CoMP cellular networks. In *Proceedings of IEEE GLOBECOMM*, December 2012.

[25] Samsung (2011) R1-112518, Discussion on CSI feedback for MU-MIMO enhancement. Samsung, Technical Report, August 2011.

[26] Motorola Mobility (2011) R1-112440, Phase 2 evaluations of joint transmission (JT) and dynamic cell selection (DCS) schemes. Motorola, Technical Report, August 2011.

[27] Fujitsu (2011) R1-112660, phase 2 performance evaluation of JT CoMP. Fujitsu, Technical Report, August 2011.

[28] Li, X., Cui, Q., Liu, Y., and Tao, X. (2012) An effective scheduling scheme for CoMP in heterogeneous scenario. In *Proceedings of IEEE Symposium on Personal, Indoor and Mobile Radio Communications (PIMRC)*, September 2012.

[29] Fang, Y. and Thompson, J. (2011) Out of group interference aware precoding for CoMP: A maximum eigenmode based approach. In *Proceedings of IEEE Symposium on Personal, Indoor and Mobile Radio Communications (PIMRC)*, September 2011.

[30] Cui, Q., Yang, S., Xu, Y., and Tao, X. (2011) An effective inter-cell interference coordination scheme for downlink CoMP in LTE-A systems. In *Proceedings of IEEE Vehicular Technology Conf. (VTC)*, September 2011.

9

Downlink CoMP: Standardization Impact

9.1 Introduction

Chapter 8 discussed the various signal processing aspects of CoMP. In this chapter the focus shifts to the various enhanced features that were introduced in the LTE standard (Release 11) to support CoMP.

9.2 Modification of Reference Signals

In LTE Release 10, CSI-RS and DMRS were designed for data transmission from a single base station. For example, in Release 10, the UE has to measure the channel from only its serving cell. To enable communication with multiple base stations, changes are required for both CSI-RS and DMRS. These are discussed in the following sections.

9.2.1 Modifications in CSI-RS

Assume that the serving cell has cell ID N_{ID}^{cell}. The serving cell semi-statically configures the UE with the number of antenna ports, subframes, and RE locations of the CSI-RS resources that it would transmit. When it transmits the CSR-RS sequence, it initializes this sequence by the scrambling seed

$$c_{init} = 2^{10}(7(n_s + 1) + l + 1)(2N_{ID}^{cell} + 1) + 2N_{ID}^{cell} + N_{CP}, \qquad (9.1)$$

where n_s, l, N_{CP} are the slot number, OFDM symbol within a slot, and the cyclic prefix parameter, respectively. Since the UE is aware of the cell ID of its serving cell, it can descramble the CSI-RS sequence.

For CoMP, the UE has to measure the channel from multiple points. Is this possible using the Release 10 CSI-RS structure? Consider a heterogeneous network with a macro- and a picocell and a representative UE. Suppose that the UE is associated with the macrocell and has to report

Heterogeneous Networks in LTE-Advanced, First Edition. Joydeep Acharya, Long Gao and Sudhanshu Gaur.
© 2014 John Wiley & Sons, Ltd. Published 2014 by John Wiley & Sons, Ltd.
Companion Wesite: www.ltehetnet.com.

channel states for both the macro and pico. If the Release 10 CSI-RS structure is maintained, the UE can only descramble CSI-RS sequences that are initialized by the macrocell ID. In order for the UE to descramble any CSI-RS sequence transmitted by the picocell, the pico has to transmit a CSI-RS sequence that has been initialized with the macrocell ID. This is in addition to the usual CSI-RS, scrambled with the picocell ID, that the pico has to transmit for UEs that are associated with itself. The picocell therefore transmits CSI-RS twice which increases the RS overheads.

Even if the pico were to transmit the CSI-RS twice, there is still another problem that can prevent a UE from effectively measuring the channels from both the macro- and picocells. Recall that the LTE standard gives flexibility to the UE to estimate the channel of a given subframe by measuring the received CSI-RS signals over multiple subframes. For a Release 10 UE, the same point transmits all the CSI-RS sequences over the multiple subframes. In the proposed method the pico transmits in some subframes instead of the macro with the same CSI-RS configuration parameters as the macro in a manner transparent to the UE. The method would provide inaccurate CSI estimates if the UE collected CSI-RS from the pico transmission subframes for estimating the macro channel and vice versa. This can be resolved by using *restricted measurement subsets* [1, section 7.2]. Restricted measurement subsets were introduced for FeICIC for the situation when the UE had to feed back CQI values for ABS and normal subframes separately. The UE could be RRC configured with two different sets of subframes (corresponding to the ABS and normal subframe patterns), so that it can measure and report two different CQI values CQI_{ABS} and CQI_{normal}, respectively.

This process does not scale beyond two nodes, but provides hints about what should be done in Release 11 to support CoMP. The first is to change the *cell-specific* nature of the scrambling ID to *UE-specific*. Accordingly, in Release 11 the seed value used to initialize the scrambling sequence was modified to

$$c_{\text{init}} = 2^{10}(7(n_s + 1) + l + 1)(2Y + 1) + 2Y + N_{\text{CP}}, \qquad (9.2)$$

where Y can take any value from the range of possible cell IDs and does not have to be the cell ID of the serving cell (or any other cell in the CoMP measurement set). The network decides the value of Y for all base stations for which the UE has to measure the channel, and informs the UE via RRC signaling.[1] This modification of the scrambling seed is also very helpful for CoMP scenario 4 as all the nodes have the same cell ID, but can now have different values of Y so that the UE can distinguish their CSI-RS signals.

Using these enhancements, the previous macro–pico CSI reporting is now possible. The UE is first configured with two CSI processes. In the subframes of the first process, the macro transmits CSI-RS with initialization parameter Y_{macro}; in the subframes of the second process, the pico transmits CSI-RS with initialization parameter Y_{pico}. The values of Y_{macro} and Y_{pico} are RRC configured to the UE. It can therefore measure and report the CSIs from both the nodes.

9.2.2   Modifications in DMRS

Similar enhancements are also required for DMRS. One of the main applications of DMRS-based transmissions is to enable MU-MIMO in Release 10 and, by natural extension,

---

[1] However if the network does not configure a value of Y, the UE assumes it to be equal to the serving cell ID. This maintains backwards compatibility, as a Release 11 UE can also be operated in Release 10 mode.

MU-JT in Release 11 CoMP. To understand the DMRS design for Release 11, let us first consider the Release 10 DMRS sequence. It is scrambled with a seed initialized by

$$c_{\text{init}} = (\lfloor n_s/2 \rfloor + 1)(2N_{\text{ID}}^{\text{cell}} + 1)2^{16} + n_{\text{SCID}}. \qquad (9.3)$$

The DMRS sequence is therefore scrambled by two parameters: the cell ID $2N_{\text{ID}}^{\text{cell}}$, which is an example of a *group-scrambling ID*, and a *scrambling sequence ID* n_{SCID} which has the value either 0 or 1. Release 10 supports up to four layers for MU-MIMO. Since the UE has to measure the precoded channel for all four layers, the base station should be able to send four DMRS sequences (one for each layer such that they can be distinguished at the UE side). There are multiple ways to achieve this [2], as follows.

1. Allocate four orthogonal DMRS ports. Orthogonal DMRS ports can be created by having non-overlapping sets of REs in frequency and time (same principle as CRS) and/or reusing the same set of REs with code division multiplexing (CDM). Using CDM gives better power control across layers and is preferred. Specifically, two non-overlapping sets of 12 REs/PRB have been proposed. In each set of 12 PRBs, two orthogonal ports are derived via CDM.
2. Allocate two orthogonal DMRS ports. For each port use two DMRS sequences that are scrambled using different values of scrambling sequence ID, n_{SCID} (0 or 1). Using this method does not give complete orthogonality when compared to the first option, but still provides *interference randomization* or *quasi-orthogonality*. In other words, a residual inter-layer interference will remain and the channel estimation accuracy would suffer. Since only two ports are used however, this has less DMRS overheads (i.e. 12 REs compared to the first option which used 24 REs).

As common in the 3GPP standardization process, multiple potential solution approaches are proposed by the member companies which are then rigorously tested. In this case, the second method was adopted for DMRS by common consensus at the end of the Release 11 evaluation process. The two antenna ports are designated antenna ports 7 and 8.

However, the following problems still remain [3].

1. In CoMP scenario 4, all cells have the same cell ID. A maximum of four UEs can therefore be scheduled in the whole network for MU-MIMO, a severe restriction. Slight performance gain may be achieved through spatial reuse where far-away UEs can be scheduled in the DMRS antenna port that uses the same n_{SCID}. However, the performance gains achievable by this method are not usually significant.
2. In CoMP scenario 3, cells have different cell IDs. The DMRS sequences between different cells are quasi-orthogonal as they have different $N_{\text{ID}}^{\text{cell}}$, but may share the same antenna port. This causes interference between the DMRS sequences at the UE for cell-edge UEs in any legacy system. However, due to the dense deployment of macro- and picocells, the problem is more severe in a heterogeneous network [4].

The basic problem arises as the group scrambling ID is fixed to the value of the cell ID, $N_{\text{ID}}^{\text{cell}}$. This limits the multiplexing flexibility of DMRS [5], as for CSI-RS. Similarly to CSI-RS, the adopted solution is to define the group scrambling ID in a UE-specific manner. The Release 11 DMRS initialization seed is therefore given by

$$c_{\text{init}} = (\lfloor n_s/2 \rfloor + 1)(2X + 1)2^{16} + n_{\text{SCID}} \qquad (9.4)$$

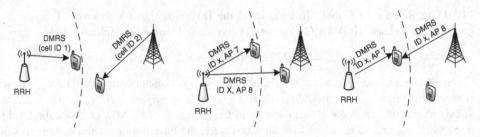

Figure 9.1   DMRS port initialization for MU-JT

where X is semi-statically configured to the UE via higher layers. Other potential approaches to UE-specific scrambling, such as defining multiple values of n_{SCID} in a UE-specific manner or introducing additional orthogonal scrambling sequences on top of the Release 10 DMRS [6], were not adopted in Release 11.

Figure 9.1 is an example to illustrate the benefits of the new DMRS sequences. In the left-most side, two cell-edge UEs are being scheduled for single-layer transmission by the macro- and the picocell in the same resource block. The DMRS sequences are quasi-orthogonal, leading to inter-stream interference. In Release 11, the two UEs can both be served by the picocell (macro is blanked) with MU-JT. This is depicted in the central figure. The DMRS sequences are initialized by the same group-scrambling ID X and the two UEs are scheduled in orthogonal ports to eliminate inter-layer interference. The right figure depicts a SU-JT transmission where one UE receives an independent data layer from each base station. In this case, the two cooperating base stations share a common DMRS group scrambling ID X for the purposes of this cooperative transmission (i.e. they can use their actual cell IDs for transmission to other UEs) and the UE is informed of the value of X semi-statically. Since the base stations use the same group-scrambling ID, the different layers can be allocated orthogonal antenna ports.

In practice, if X is chosen from a large set of potential values and the network has to RRC configure the exact value of X for each UE, the RRC overhead would increase significantly. In Release 11 two default values of X, namely $X(0)$ and $X(1)$, are configured for each UE. Being default values they do not have to be transmitted via RRC signaling every time the UE has to use one of them. RRC configuration is used only if the network feels that the UE needs to be configured with additional values of X. The important question is: how does the network inform the UE about whether to use $X(0)$ or $X(1)$? RRC or extra DCI signaling has been avoided for this purpose by mandating UE behavior to make it choose $X(0)$ if $n_{SCID} = 0$ and $X(1)$ if $n_{SCID} = 1$.

The only remaining question is: how to choose the values of $X(0)$ and $X(1)$? The values can relate to possible values of Y used for CSI-RS scrambling [7]. Ultimately, this has been left to proprietary implementation. If these values are not specified, the UE assumes them to be equal to the cell ID of the serving cell. The above discussion serves as a reminder of the constant trade-off between performance of an algorithm and simplification of its implementation complexity that is so prevalent in standardization design.

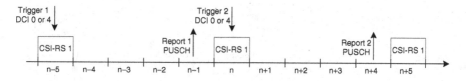

Figure 9.2   Illustration of CSI processes, CSI trigger, and feedback for Release 10

9.3   CSI Processes

In this section, we discuss the CSI reporting procedure for multiple CSI reports. To understand the details, we first recall the CSI reporting procedure in Release 10. For periodic feedback, different kinds of reports (e.g. one for PMI/RI and another for CQI) and the subframes where feedback should occur are RRC configured to the UE. The major issue in periodic feedback is collision handling between the different feedback reports which are configured with different periodicities and can therefore collide in a given subframe. For aperiodic feedback, the base station asks the UE to transmit CSI via dynamic signaling. To implement this, the base station transmits an *aperiodic trigger* to the UE contained in DCI formats 0 or 4 which includes a CSI request field. The length of this field is 1 or 2 bits. The length is 2 bits when the UE is capable of carrier aggregation, in which case the bits are used to trigger feedback for different sets of component carriers. The process is depicted in Figure 9.2.

Note that there are three different kinds of time instances in Figure 9.2:

1. when the CSI resources (in this case CSI-RS) occur;
2. when the CSI request triggers occur; and
3. when the UE feeds back the CSI report.

For Release 10, the CSI request trigger occurs in a subframe that contains the CSI resource. To understand the reason for this, recall that in Release 10 the interference is assumed to be measured based on CRS. Since the CRS occurs in each subframe, the interference measurement can also be performed in each subframe. By sending the trigger on a subframe that contains the CSI-RS, both channel and interference measurement subframes will therefore be aligned. The UE sends the CSI report four subframes after it has received the trigger. This means that a Release 10 UE has up to three subframes to process the channel estimation values into a CSI report.

The situation changes when the UE has to feed back multiple CSI reports that correspond to different CSI resources in Release 11. The occurrence of multiple CSI resources which lead to multiple CSI reporting is captured by the term *CSI Process* [8]. A CSI process is a time series of subframes where the CSI-RS and IMR corresponding to a given transmission hypothesis occur and also the time series of subframes where the CSI is fed back. The term *process* in CSI process is somewhat inspired by the expression 'stochastic processes', which describes the time evolution of random sample paths.

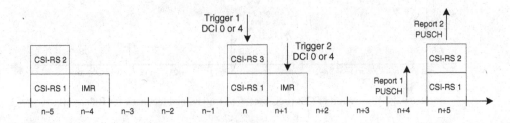

Figure 9.3 Illustration of CSI processes, CSI trigger, and feedback in Release 11

More formally, a CSI process is defined as follows.

Definition 9.3.1 *A CSI process is RRC configured to the UE by the association of:*

- *one non-zero-power CSI-RS resource from the CoMP measurement set;*
- *one interference measurement resource (IMR);*
- *one CSI reporting mode (PUCCH or PUSCH); and*
- *other feedback-related parameters such as codebook subset restriction.*

Returning to the macro–pico network in Table 8.2 of the previous chapter, it can be said that the UE had been configured with four CSI processes in that example.

Figure 9.3 shows an example of three CSI processes. CSI process k ($k = 1, 2, 3$) is defined by the combination of CSI-RS$_k$ and IMR. The three CSI processes share a common IMR, which is the case when the interference to be measured is the same. In Table 8.2 for example, the CSI processes corresponding to hypotheses 1 and 3 had the same out-of-CoMP set interference to measure. The periodicities of these processes can be different. For example, the first process occurs with periodicity 5 while that of the second and third occur with periodicity 10.

Figure 9.3 shows some important changes from Release 10 CSI configuration. Firstly, the CSI resource is now split into a CSI-RS and IMR, which can occur in different subframes. For aperiodic reporting, the CSI request trigger need not be in the same subframe as the CSI-RS and/or IMR. However, the CSI report is still transmitted four subframes after receiving the CSI request trigger.

In Release 11, the UE has multiple CSI resources for which it can report back CSI. At any time instant, the base station may need reports for only a subset of them. Feedback of multiple sets of CSI processes is similar to feedback of CSI values for multiple sets of component carriers. The scope of the two CSI request bits of DCI formats 0 and 4 has been expanded to include feedback request from multiple sets of CSI processes. A possible implementation is shown in Table 9.1. As seen in the table, the UE is RRC configured by the network to feed back a certain set of CSI process(es) for each possible value of the CSI request field bits. For example, consider the macro–pico deployment situation and the various transmission hypotheses shown in Table 8.2 of the previous chapter. The network may configure the UE to transmit CSI corresponding to hypotheses 1 and 3 for value '01', hypotheses 2 and 4 for '10' and hypotheses 1–4 for value '11'. The base station therefore implements the trigger by a combination of RRC (semi-static) and DCI (dynamic) signaling.

CA and CoMP operations are simultaneously permissible in Release 11. For a UE that is capable of both these operations, the CSI request field indicates feedback of CSI corresponding

Table 9.1 Example of aperiodic CSI triggering for CoMP and CA by using a 2 bit CSI request field

Value of CSI request field	Description
'00'	No aperiodic CSI is triggered
'01'	A report is triggered for a set of CSI process(es) configured by higher layers for serving cell c
'10'	A report is triggered for a first set of CSI process(es) configured by higher layers
'11'	A report is triggered for a second set of CSI process(es) configured by higher layers

to different transmission hypotheses for different component carriers. Note from Table 9.1 that the codepoint '01' is used to trigger CSI processes for the serving cell (it could be the primary component carrier). To enable simultaneous CA and CoMP feedback, a CSI process is identified by a CSI process index and a serving cell index.

9.3.1 UE Processing Complexity and CSI Reference Resources

In Release 10, the worst-case CSI feedback processing occurs for aperiodic CSI reporting, especially with eight transmit antenna dual codebook structure when the UE has to compute two PMIs. Even for four transmit antennae, aperiodic reporting contributes significantly to UE processing complexity [9]. This section discusses UE processing complexity for multiple CSI resource feedback and features introduced in Release 11 to limit this complexity to acceptable levels.

Two instances of aperiodic triggers are shown in Figure 9.3, which occur in subframes n and $n + 1$. Consider the case when the aperiodic trigger requests the UE to report CSI for all three resources. When the UE receives the CSI trigger in subframe n, it computes CSI based on CSI-RS information in subframe n and IMR in subframe $n - 4$. On receiving a trigger in the next subframe, what should it do? If no UE behavior is specified in the standard and it is left to proprietary UE implementation, then the UE may recompute all the CSI reports based on the new IMR in subframe $n + 1$. If a serial processing model is adopted (i.e. UE finishes CSI computing for all processes specified by a trigger and then starts processing for the next trigger), there is only one subframe for the UE to compute all CSI processes [10]. The UE processing therefore becomes highly complex.

The solution of this problem is to ensure that the UE does not need to compute CSI frequently. It would be desirable to have a design in which there are always at least N subframes between the instances where the UE starts CSI computation. There are at least two ways to realize this design; the first is depicted in Figure 9.4. Here the CSI-RS and IMRs belonging to different CSI processes are configured in the same subframe. Since the minimum periodicity of CSI-RS and IMR is 5 subframes, this means that the UE will always have at least 3 subframes to process the CSI reports. For example, when the first CSI trigger arrives at subframe n, the CSI-RS and IMR values from subframe $n - 4$ are used to compute the CSI reports which would then be fed back in subframe $n + 4$. The UE effectively has 7 subframes to process this report. When the

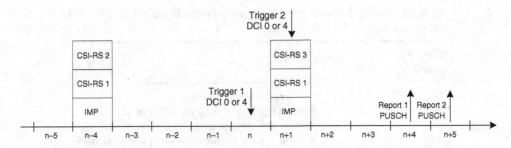

Figure 9.4 Example configuration of restricted CSI-RS and IMR configuration so that the UE has to compute CSI periodically

second CSI trigger arrives in subframe $n + 1$, the CSI-RS and IMR values in subframe $n + 1$ are used to compute a new CSI report which is fed back in subframe $n + 5$. The UE therefore has 3 subframes to process this report.

Note that the explanation about how there are 7 subframes for calculating the first CSI report may be slightly misleading. It is true that there are 7 subframes between the subframe where the CSI-RS and IMR occur and the subframe where the corresponding report is fed back. But 4 of these subframes occur before the UE is even aware of an aperiodic trigger. In this case, should subframes $n - 4$ to $n - 1$ be counted towards CSI processing time? The answer is in the affirmative. Although the UE receives the trigger later it may still use these 4 subframes for channel pre-processing purposes, which needs to be evaluated continuously regardless of actual reporting [11].

This method reduces UE complexity by specifying network behavior in terms of RS configuration. It is a restrictive design however as it forces the CSI-RSs and IMRs to be in the same subframe. The second approach is to achieve the same goal of periodic CSI computation by specifying UE behavior. This is done by introducing the notion of *CSI Reference Resources* which are subframes where the UE starts CSI processing. Release 11 specifies that the CSI Reference Resources will occur periodically every N subframes with $N = 5$ for FDD. This is shown in Figure 9.5. In this example, the reference resource is chosen to the subframe where the IMR occurs. When the first trigger arrives in subframe n, the valid CSI reference resource is subframe $n - 4$.[2]

The CSI processing (or pre-processing) is essentially done in a more periodic fashion, even although the trigger and reporting instances occur aperiodically. Although the examples in this section have discussed aperiodic feedback, CSI reference resources are also defined for the purposes of periodic feedback.

9.3.2 Inheritance and Reference Processes

Feedback for a CSI process includes parameters such as PMI/RI and/or CQI, but there are usually dependencies between the parameters that are being fed back from different processes.

---

[2] The standard states that, for CSI reporting in subframe N, the CSI reference resource of the CSI process is the first valid CSI reference resource occurring on or prior to subframe $N - x$ where $x = 5$ for FDD. In our example, the first reporting instance is therefore at subframe $n + 4$ and the relevant CSI reference resource should occur on or before subframe $n - 1$. As seen from the figure, this CSI reference resource is subframe $n - 4$.

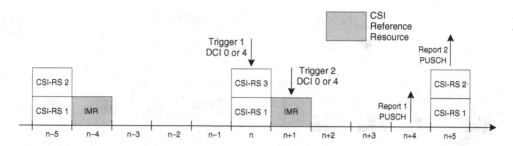

Figure 9.5   Example of using periodic CSI reference resource subframes so that the UE has to compute CSI periodically

As seen in Example 8.4.2 from the previous chapter, for hypotheses 1–4 a valid precoder selection is that $\mathbf{W}_2 = \mathbf{W}_1$ and $\mathbf{W}_4 = \mathbf{W}_3$. In this case, the UE should feed back only one of $\mathbf{W}_2$ and $\mathbf{W}_1$ and one of $\mathbf{W}_4$ and $\mathbf{W}_3$. To ensure this, the network can configure CSI process 1 to feed back CQI and PMI/RI and a second CSI process 2 to feed back only CQI assuming the same PMI/RI as the first process. An as example, let CSI process 1 correspond to hypothesis 1 and CSI process 2 correspond to hypothesis 2. CSI process 2 is said to *inherit* the PMI/RI from CSI process 1, and CSI process 1 is said to be a *reference process* for CSI process 2.

Rank inheritance is important for enabling CoMP [12]. In Example 8.4.2, CSI processes 1 and 3 feed back per-point precoders $\mathbf{W}_{\text{macro}}$ and $\mathbf{W}_{\text{pico}}$, respectively. If the network wants to perform JT CoMP, it has to estimate the JT precoder $\mathbf{W}_{\text{JT}}$ from the per-point precoders. It would be helpful if the reported RIs of $\mathbf{W}_{\text{macro}}$ and $\mathbf{W}_{\text{pico}}$ are the same, as then it becomes easier to combine them and obtain $\mathbf{W}_{\text{JT}}$. If the RIs of $\mathbf{W}_{\text{macro}}$ and $\mathbf{W}_{\text{pico}}$ are different, combining them is not straightforward. Also, it is easier to implement frequency-selective scheduling of different CoMP transmission schemes (such as frequency-selective DPS, where the UE receives data from different base stations in different sub-bands), if the CSI reports from the base stations have the same transmission rank. A rank indicator (RI) reference process has therefore been standardized for aperiodic feedback. The RI of a process can be configured to inherit its value from the RI reported in the same subframe of the RI reference process.

9.4   PDSCH Rate Matching

In Releases 8–10, the UE knows the CRS locations of the serving cell. It also knows that the PDSCH is rate-matched around the CRS.[3] In Release 11 CoMP, the set of point(s) transmitting PDSCH can change dynamically. Each point may have a different cell ID and hence a different locations of CRS. An example is depicted in Figure 9.6. Consider a UE which is associated with the macro but is served by the pico in a subframe via DPS. Since the CoMP operation is transparent to the UE, it does not know the transmitting base station. However, the UE needs to know the CRS REs of the transmitting cell to know which REs carry PDSCH. The only way the UE can know this is for the network to signal the CRS port information (number of CRS ports and their location) of the transmitting point [13]. This has to be transmitted by dynamic DCI signaling as the transmitting point may change in each subframe. In case of JT,

---

[3] UE knows that the base station will not transmit PDSCH in the REs corresponding to CRS REs.

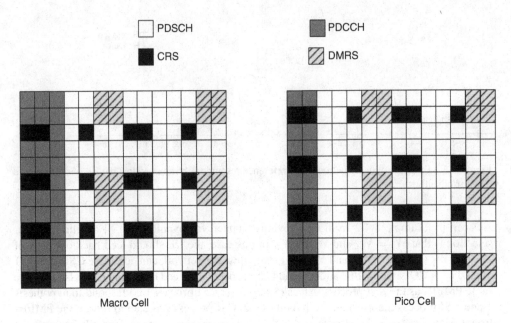

Figure 9.6 Example of PRBs transmitted by a macrocell and an adjacent picocell. If they have different cell IDs, the CRS locations of the two cells are shifted

the network may decide to rate match around the CRS locations of both macro- and picocells (i.e. each cell does not transmit PDSCH in REs corresponding to the CRS locations of both cells). The CRS ports around which PDSCH is rate matched has to be transmitted to the UE.

The starting symbol for the PDSCH region can also be different in different cells. As shown in Figure 9.6, the macrocell has three OFDM symbols in the PDCCH region but the picocell has two. In Releases 8–10, the UE is used to decode PCFICH to determine the length of the PDCCH region of the serving cell. This may not be possible for other points in the CoMP cooperating set, however. The network can ensure that all cells that transmit to a UE via CoMP have the same PDSCH starting region. This will place unnecessary restrictions, especially when ePDCCH is employed in one cell for which the PDSCH can start from the very fist symbol.

The zero-power CSI-RS configuration also changes from cell to cell. Each cell can configure separate patterns for zero-power CSI-RS to mute interference for some other cells' CSI-RS transmission. Additionally, each cell has its own IMR pattern also configured from the set of zero-power CSI-RS REs. Overall, the UE needs to know all REs of the serving cell that contain zero-power CSI-RS to rate match the PDSCH around them.

Another factor that affects the number of available REs for PDSCH is the MBSFN subframe configuration of the cells in the CoMP set. In Release 10, each cell has its own MBSFN configuration which is signaled to its UEs by SIB-2. In Release 11 CoMP, a UE may be served by cells having different MBSFN configurations as shown in Figure 9.7. Since MBSFN subframes do not have CRS REs, the number of REs available for PDSCH in a subframe will change depending on which cell transmits to the UE.

There are multiple ways to inform the UE about the PDSCH rate matching. The network can implement DPS and JT only in MBSFN subframes to avoid the need for CRS RE information [14]. However, this will restrict the opportunities for CoMP. The network can use a common

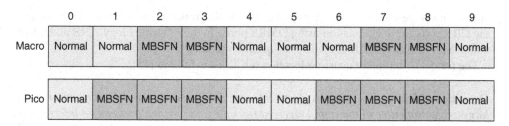

Figure 9.7   Example of a CoMP cooperating set with a macro- and a picocell where the two cells have different MBSFN configurations

CRS configuration for both cells (feasible in CoMP scenario 4, for example). UE behavior may be specified to define only two states of CRS RE locations for rate matching: either that of its serving cell or a common configuration covering the CRS locations of all cells in the CoMP set. All these approaches were evaluated and discussed during the Release 11 standardization phase. Eventually it was decided to define a new DCI format for CoMP to inform the UE about the CRS RE information needed for PDSCH rate matching. The details of this new DCI format are discussed in Section 9.6.

9.5   Quasi-Co-Location of Antenna Ports

We have discussed several aspects of CoMP including MIMO precoding, MCS determination, and feedback structure. There are other important communication technology challenges that have to be addressed before the theoretical gains of CoMP can be realized, however. These challenges arise due to that basic nature of CoMP: transmission of signals from multiple points that are not co-located.

To understand this, consider the example of a UE trying to decode its received PDSCH. To do so, it has to first estimate the instantaneous channel using the associated DMRS. The UE needs knowledge of certain large-scale and wideband properties of the channel before it can estimate the instantaneous channel realization in the allocated PRBs. An example of these large-scale properties are power delay profile, delay spread and Doppler shift of the channel. The UE needs to know these values to initialize the Wiener filter to be used for DMRS-based channel estimation. How does the UE obtain these values? One obvious answer is that the UE tries to estimate both the large-scale properties of the channel and its instantaneous realization from the received DMRS.

However, DMRS for a UE occur only in the subframes and PRBs allocated for that UE's transmission; it is therefore sparse in time and frequency and difficult to estimate the large-scale and wideband properties of the channel using DMRS alone. What can the UE do? Suppose that the large-scale channel properties from a second set of antenna ports are the same as those observed in the DMRS antenna ports. The UE can use the second set of ports to estimate the large-scale properties, if it is easier to estimate these properties using the second set of antenna ports. For instance, in Release 10, the UE uses CRS to estimate the large-scale properties of the channel. CRS is transmitted frequently and thus the estimates of these channel properties will be accurate. Since CRS and DMRS are transmitted from the same set of physical antennae, the large-scale properties of the channel seen in both sets of antenna ports are same. The UE can use the large-scale properties of the channel derived from the CRS ports for demodulation using DMRS ports. In 3GPP terminology, CRS and DMRS antenna ports in Release 10 are said to be *quasi-co-located*.

The formal definition of quasi-co-location (QCL) is as follows [15].

Definition 9.5.1 *If two antenna ports are quasi-co-located, the UE may assume that large-scale properties of the signal received from the first antenna port can be inferred from the signal received from the other antenna port. The large-scale properties mentioned in the above definition consist of some or all of the following:*

- *delay spread;*
- *Doppler spread;*
- *frequency shift;*
- *average received power; and*
- *received timing.*

From this definition, the UE cannot assume that non-quasi-co-located antenna ports have the same large-scale channel properties. It therefore has to perform independent processing, such as timing acquisition and tracking, frequency offset estimation and compensation, channel estimation, and Doppler estimation for each configured non-quasi-co-located antenna port. For example, it is straightforward to see that in Release 11 CoMP scenario 4, the antenna ports of CRS and DMRS are not co-located. CRS is transmitted in an SFN manner from all antenna ports and thus represents the composite channel while DMRS characterizes the per-point channel between a point and the UE.

The downlink demodulation and decoding depends on the UE assumptions about quasi-co-location behavior between the different antenna ports as the channel estimation, Doppler compensation, CSI reporting, etc. are all affected [16].

The notion of quasi-co-location was introduced to understand how the large-scale properties of the channel can be estimated by a UE when it receives signals from geographically separated antenna ports. For this situation, some important questions arise as follows.

1. For a given group of transmitting points, each with multiple antenna ports, which of these antenna ports are quasi-co-located?
2. For a given UE which receives signals over multiple antenna ports, which of these antenna ports may be assumed by the UE to be quasi-co-located?
3. If the UE makes a certain assumption about the quasi-co-location between a set of antenna ports, how does this affect CoMP transmission at the base station?

Note that although questions (1) and (2) seem identical, they are not. To understand this, consider the example of a four-port CSI-RS resource. It can be transmitted from four co-located physical antennae or it can be an aggregated resource from two base stations each having two physical antenna ports. Clearly, in the first case all ports are quasi-co-located while in the second case ports $\{1, 2\}$ and $\{3, 4\}$ are quasi-co-located between themselves but ports $\{1, 2\}$ need not be quasi-co-located with $\{3, 4\}$. This is the answer to question (1).

To answer question (2), note that the UE is not aware if the CSI-RS resource is aggregated or per point when it receives it. The network can inform the UE about what ports it should assume to be quasi-co-located. However, this increases RRC signaling to the UE. The UE can therefore assume that all ports in a received CSI-RS resource are always quasi-co-located and ports belonging to different CSI-RS resources are not quasi-co-located.

This also provides an answer for the third question. For this UE quasi-co-location assumption, the network will not configure an aggregated CSI-RS resource; it knows that if it does so, the UE will not be able to process the CSI-RS resource reliably. Alternatively, the network can try to find two base stations having similar large-scale properties (i.e. they are still quasi-co-located despite not being physically co-located) and configure an aggregated resource.

The remainder of this section provides answers to these questions in detail. Before doing so, two characteristics about quasi-co-location should be stressed. Firstly, if two antenna ports correspond to co-located physical antenna terminals, they will also be quasi-co-located. However, if two sets of antennae are not physically co-located, it does not automatically imply that their large-scale channel properties are not correlated. For CoMP operation, it is possible for the network to choose two distributed RRHs that have strongly correlated large-scale channels to a UE, i.e. two RRHs that are quasi-co-located.

Secondly, note the phrase 'some or all of' in Definition 9.5.1. This means that if a UE assumes that two antenna ports are quasi-co-located for performing an operation such as channel estimation, it may only need some of the large-scale channel properties to be correlated. To understand this, note that the large-scale properties of a channel can be loosely grouped based on how they are used by the UE as follows [17].

1. **Average received power:** The UE performs RSRP measurements over multiple antenna ports. If the antenna ports are quasi-co-located, it can achieve a better estimate of RSRP by averaging over them.
2. **Frequency shift and received timing:** The UE needs to estimate the received timing and correct the frequency offset error of the received signal before passing it to the FFT. The UE can obtain values of these parameters for one port and use them for another quasi-co-located port.
3. **Delay spread and Doppler shift:** These values are used to initialize the Wiener filter for channel estimation. The UE can obtain values of these parameters from one port and use them for another quasi-co-located port.

9.5.1 Quasi-Co-Location Between the Same Antenna Ports

In this section quasi-co-location assumptions between sets of CRS ports, sets of CSI-RS ports, and sets of DMRS ports is discussed.

9.5.1.1 QCL between CRS Ports

Up to Release 11, a UE takes RSRP measurements using CRS. CRS port 0 is always used for RSRP measurements and if CRS port 1 can be detected by the UE, it can also be used. Although it has not been explicitly stated in the standards, CRS ports 0 and 1 are assumed to be quasi-co-located by the UE; this is because there is no relevant UE behavior for the assumption that the CRS ports are not co-located. Even for a Release 11 transmission, the UE may assume that all CRS ports are quasi-co-located. It is also worthwhile mentioning that CRS port 0 and the PSS/SSS of the serving cell are assumed to be the reference for all front-end processing by the UE. The UE may therefore assume that PSS/SSS ports and CRS port 0 are quasi-co-located.

9.5.1.2 QCL·between CSI-RS Ports

Quasi-co-location between different ports of a CSI-RS resource was discussed in the example of a four-port CSI-RS resource at the beginning of this section. In Release 11 the UE assumes that all CSI-RS ports belonging to one CSI-RS resource are quasi-co-located and CSI-RS ports belonging to different CSI-RS resources are not quasi-co-located. Although this does not necessarily preclude the configuration of an aggregated CSI resource, it significantly limits its possibility.

9.5.1.3 QCL between DMRS Ports

For SU/MU-MIMO and non-frequency-selective DPS, the PDSCH and DMRS in all the allocated PRBs come from one base station. All DMRS ports will therefore be quasi-co-located. In frequency-selective DPS, PDSCH in different PRBs can be transmitted by different base stations. In LTE, a *Precoding Resource Group* (PRG) is a group of PRBs for which the UE can assume that the same precoder had been used by the eNodeB. For frequency-selective DPS, a logical conclusion is that a UE should assume that all DMRS ports are quasi-co-located for a given PRG and not quasi-co-located across PRGs.

This assumption will lead to increased UE processing complexity during the demodulation process. UEs will need knowledge of the reference timing for each PRG to compensate the received signal accordingly. After performance evaluations and lengthy discussions, it has been agreed that the gains of frequency-selective DPS were not significant. UE behavior based on PRG level quasi-co-location assumptions is therefore excluded from Release 11 CoMP. This does not directly preclude the use of frequency- selective DPS, but limits its possibility for all practical purposes.

It must be clarified that, unlike frequency-selective DPS, the per-subframe-level quasi-co-location assumption among all DMRS ports does not hamper the implementation of JT [18]. In JT, the channels of the cooperating base stations can have different large-scale properties but, since they transmit in all PRBs simultaneously, the UE perceives a combined channel whose combined large-scale properties stay the same in all PRBs of the subframe. The subframe level quasi-co-location assumption of DMRS is usually good enough for data demodulation in JT. If the large-scale parameters (especially received timings) of the base stations are very different this could be a problem, but this can be fixed by the network choosing base stations with similar received timings.

9.5.2 *Quasi-Co-Location Between Different Antenna Ports*

In this section the quasi-co-location assumptions between different sets of antenna ports is discussed. The motivation is that a set of antenna ports (such as DMRS) may not be able to track the large-scale channel properties by themselves and will need the support of other ports.

9.5.2.1 QCL between CSI-RS and CRS Ports

Recall that configuration of aggregated CSI-RS resources has been limited due to reasons related to QCL. But does this mean that the alternative, which is per-point CSI-RS configuration, has no need for any QCL assumptions with another set of ports for reliable channel estimation? A possible QCL behavior for CSI-RS could be with the CRS ports from the corresponding point. Another way to pose the same question is: does the network need to tell the UE about the cell ID of the base station transmitting the CSI-RS for effective channel estimation?

To answer this, let us look into the large-scale properties individually and see if the CSI-RS ports can track them without the help of the corresponding CRS. First is the *received timing*. It is assumed that a CoMP-capable UE may operate with a single FFT timing per receiving antenna port to perform all CSI- and demodulation-related operations. This implies that the received timings of different per-point CSI-RS resources cannot be very different. It has therefore been decided that a Release 11 UE is expected to receive CSI-RS resources within a time window of $[-3, +3]$ μs with respect to the reference timing, where the reference timing is derived from PSS/SSS and CRS port 0 of the serving cell. The phrase *expected to receive* means that the network will ensure that the different CSI-RS resources that are configured to the UE have the required timing property. For this assumption, system level evaluations have demonstrated that time tracking using CSI-RS is mostly robust in fading channels.

The next large-scale property is *average received power* which is needed for RSRP measurements. For CSI-RS, RAN4 studies have shown that no significant RSRP measurement degradation is noticed by offsetting CSI-RS reception in the range $[-3, +3]$ μs as compared to the reference FFT timing. Such results are based on the assumptions that CSI-RS are not quasi-co-located with other RS types. It is also possible for a UE to estimate the power-delay-profile or delay spread for each CSI-RS resource independently of any CRS. However, frequency offset estimation performance using only CSI-RS is not satisfactory under certain channel conditions [19].

There is therefore merit in the argument that, for each configured CSI-RS resource, the network should inform the UE about which CRS ports it is quasi-co-located with. Since UEs would have the timing and frequency offset information of each cell through continuous cell search and monitoring, it would be helpful if each configured CSI-RS resource in the CoMP measurement set was linked to an associated cell ID. The UE could then use the reference time of the associated cell for initial timing acquisition and subsequent tracking of time and frequency variations of the associated CSI-RS resource in the CoMP measurement set [20].

There are also arguments against introducing CSI-RS to CRS quasi-co-location, however. The UE complexity increases with this new behavior. The actual error in CSI estimation might not be significant even if frequency offset error is introduced; this is because PMI/RI and CQI feedback is quantized which is invariant to a range of frequency-offset errors. After much deliberations, Release 11 did not introduce a UE behavior where it assumes that ports of a CSI-RS resource are quasi-co-located with ports of a CRS resource.

9.5.2.2 QCL between DMRS and CSI-RS Ports

For reliable demodulation, it has been demonstrated that DMRS alone is not sufficient for tracking the large-scale properties of the channel; DMRS ports need additional information from other ports. Since all DMRS ports will be sent from the same base station with high probability, these ports will be quasi-co-located with the set of CSI-RS ports that were transmitted from that base station. In Release 11, for a given set of DMRS ports the network can inform the CSI-RS resource of which ports are quasi-co-located with the DMRS ports.

9.5.2.3   QCL Between CRS and DMRS Ports

Finally, we mention the issue of CRS to DMRS quasi-co-location. Such UE behavior is needed to improve the demodulation process. However, since this is already taken care of by DMRS to CSI-RS quasi-co-location information, additional CRS to DMRS quasi-co-location information is not needed.

To summarize, the following quasi-co-location assumptions for PDSCH were adopted in Release 11 CoMP [1, section 7.1.10].

1. A UE configured in transmission modes 1–10 for a serving cell may assume the CRS antenna ports of the serving cell are quasi-co-located. Transmission mode 10 is a new transmission mode that has been defined for CoMP and is discussed in the next section.
2. A UE configured in transmission modes 8–10 for a serving cell may assume the DMRS antenna ports of the serving cell are quasi-co-located.
3. A UE configured in transmission modes 1–9 for a serving cell may assume the CRS, DMRS, and CSI-RS antenna ports of the serving cell are quasi-co-located.
4. A UE configured in transmission mode 10 for a serving cell is configured with one of two quasi-co-location types for the serving cell by RRC signaling to decode PDSCH, according to the transmission scheme associated with the DMRS antenna ports.
   - **Type A:** The UE may assume the CRS, DMRS, and CSI-RS antenna ports of a serving cell are quasi-co-located.
   - **Type B:** The UE may assume the CSI-RS antenna ports corresponding to the CSI-RS resource configuration identified by the higher-layer signaling and the DMRS antenna ports associated with the PDSCH are quasi-co-located.

Type B behavior has been described in this section. Type A corresponds to the default case when the UE may assume that all ports are quasi-co-located. The network informs the UE via RRC signaling what type it should assume.

9.6   New Transmission Mode and DCI Format

The preceding sections have discussed new feedback and demodulation procedures for a UE to enable CoMP. To consolidate these new types of UE behavior, a new transmission mode called transmission mode 10 has been defined for CoMP. In transmission mode 10, the UE expects to receive requests for multiple CSI feedback such as the 2-bit trigger in DCI format 4. It also expects to be informed about the rate matching and MBSFN configuration of the serving cell and which quasi-co-location behavior it should assume (type A or type B). To signal these items of information, a new DCI format called DCI format 2D was introduced for a UE in transmission mode 10. DCI format 2D is based on DCI format 2C with the addition of 2 extra bits for conveying the rate matching and quasi-co-location assumptions. These 2 bits are also called PQI (PDSCH Rate Matching and Quasi-Co-Location Indicator) bits.

PQI bits convey information about the following parameters:

1. a cell's CRS position (number of ports and frequency shift);
2. a cell's MBSFN subframe configuration;
3. a zero-power CSI-RS configuration;
4. a value of PDSCH starting symbol; and
5. a CSI-RS resource index for DMRS quasi-co-location.

Each of the four possible states defined by the 2 PQI bits correspond to a specific set containing values for each of the above parameters. The PQI bits dynamically inform the UE which of the four states it should assume. These sets themselves can be changed semi-statically via RRC signaling. This mixture of semi-static and dynamic signaling was also seen for CSI process feedback in Section 9.3.

The discussions in the preceding sections may give the impression that standardization support of Release 11 limits the extent of CoMP. Indeed, frequency-selective DPS and aggregated CSI-RS resource are some of the features whose implementation is limited by the Release 11 procedures. However, it should be appreciated that standardization occurs slowly in small incremental steps; progress is surely made to incorporate new wireless technologies. Indeed, although CoMP was discussed during earlier releases, it was not included for the first time until Release 11. Future 3GPP standards will have more features to support CoMP. Some of the topics mentioned here (aggregated CSI resource and frequency-selective DPS) may be explicitly included.

9.7 Backhaul Support for CoMP

As discussed before, implementation of CoMP requires different types of data and signaling exchange between the cooperating base stations over the backhaul connecting them. The properties of a backhaul link such as delay and capacity depend on the transmission medium. Some typical values were listed in Table 2.1 in Chapter 2. The limitations of practical backhaul links affect the performance of CoMP. For example, if the backhaul link has finite capacity, this affects the performance of JT as it needs the base stations to exchange huge amounts of data for all the UEs. The delay of the backhaul link can affect CoMP algorithms such as inter-cell CS/CB that need rapid exchange of precoders between the cooperating base stations. Initial 3GPP evaluations provide insight into the performance loss due to non-ideal backhaul [21]. These are depicted in Figure 9.8 for CS/CB and in Figure 9.9 for JT. As seen in the figures, there is significant loss for higher delay. As expected the performance of JT suffers more as it needs more data transfer than CS/CB.

These figures emphasize the need to ensure CoMP efficiency for a practical backhaul. To achieve this, CoMP algorithms have to be redesigned by taking into account the backhaul constraints explicitly. For example, during the cooperating set formation, the network can predict the *feasibility* of a particular cooperating set by checking the status of the backhaul network *a priori* and then determining which base stations would be able to transmit data over it efficiently [22]. Alternatively, CoMP algorithms may be designed on the basis of long-term channel properties rather than short-term channel estimates [23, 24]. CoMP signal processing with non-ideal backhaul is still very much in its initial phases; current research is expected to bring about more robust CoMP algorithms in the future.

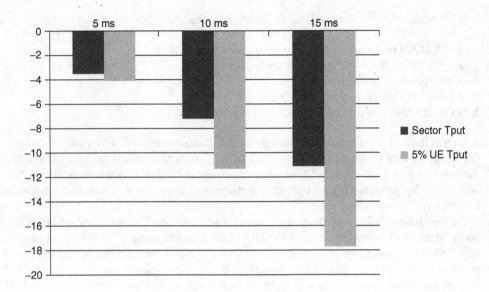

Figure 9.8 Performance degradation of CS/CB due to backhaul delay compared to a zero-delay backhaul

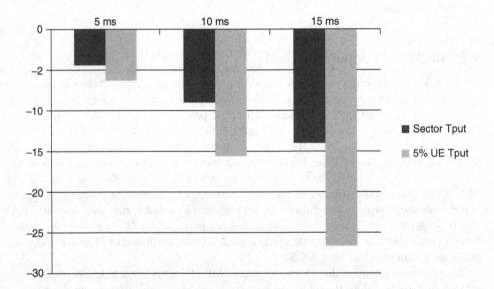

Figure 9.9 Performance degradation of JT due to backhaul delay compared to a zero-delay backhaul

9.8   Summary

The previous two chapters provided a detailed overview of Coordinated Multi-Point transmission/reception technology in 3GPP which enables different base stations in a heterogeneous network to coordinate and transmit data to UEs. Performance enhancement is possible, especially for the cell-edge UEs. However, several practical issues such as differences in the large-scale properties of the downlink channels from the cooperating base stations, multi-channel estimation, and CSI feedback pose challenges to successful CoMP implementation. Extensive work has been done at all levels including signal processing, signaling optimization, and field trials to overcome these challenges. It is no surprise that CoMP discussions which were integral to 3GPP activities right from the beginning of the Release 10 standardization phase were adopted only by the end of Release 11 activities.

Moving ahead, heterogeneous networks will become much more dense with UE traffic consuming more bandwidth than before. The UE traffic will be real-time and exhibit sharp spatio-temporal variations. These characteristics pose new and interesting challenges in the design of an efficient CoMP solution. CoMP technology has to be smart and quickly adapt to the varying network conditions. CoMP has to integrate with other network technologies, such as dynamic TDD and three-dimensional MIMO, and a holistic approach towards interference management has to evolve. Much of the work in this area is currently ongoing.

References

[1] 3GPP (2013) 3GPP TS 36.213, Evolved Universal Terrestrial Radio Access (E-UTRA), physical layer procedures. Third Generation Partnerhsip Project, Technical Report.

[2] NEC Group (2010) R1-103056, On the use of DM RS ports, scrambling sequences for MU-MIMO. NEC Group, Technical Report, May 2010.

[3] CATT (2011) R1-112960, DL Reference signal enhancement for CoMP transmission. CATT, Technical Report, October 2011.

[4] ZTE (2011) R1-113014, Evaluation on necessity of DMRS enhancement under CoMP scenarios. ZTE, Technical Report, October 2011.

[5] Ericsson (2011) R1-113355, Downlink reference signals for enhanced multiplexing. Ericsson, Technical Report, October 2011.

[6] Alcatel-Lucent, Alcatel-Lucent Shanghai Bell (2011) R1-113320, Downlink reference signal requirements in support of CoMP. Technical Report, October 2011.

[7] KDDI Corporation (2012) R1-123237, Discussions on default values of x(0) and x(1) for DM-RS sequence initialization. KDDI Corporation, Technical Report, August 2012.

[8] Huawei, HiSilicon (2012) R1-121946, CSI feedback modes for CoMP. Huawei, Technical Report, May 2012.

[9] Renesas (2012) R1-122350, On CSI feedback processing complexity in CoMP. Renesas, Technical Report, May 2012.

[10] QualComm (2012) R1-123683, Remaining details of the size of the CoMP measurement set. QualComm, Technical Report, August 2012.

[11] Ericsson (2012) R1-123825, Limiting the peak CSI processing load. Ericsson, Technical Report, August 2012.

[12] Ericsson (2012) R1-122836, RI and PMI sharing between multiple CSI processes. Ericsson, Technical Report, May 2012.

[13] Samsung (2012) R1-122241, PDSCH RE mapping for CoMP among multiple cells. Samsung, Technical Report, May 2012.

[14] QualComm (2012) R1-123691, Downlink control signaling in support of downlink CoMP. Qual-Comm, Technical Report, August 2012.

[15] Alcatel Lucent, Alcatel Lucent Shanghai Bell (2012) R1-122458, Further discussion of quasi co located antenna ports. Technical Report, May 2012.

[16] Ericsson (2012) R4-120679, Geographically separated antenna and impact on CSI estimation. Ericsson, Technical Report, February 2012.

[17] Huawei, HiSilicon (2012) R1-122512, Discussion on antenna ports colocation. Huawei, Technical Report, May 2012.

[18] Ericsson (2012) R1-124547, Outstanding issues for antenna ports quasi co location. Ericsson, Technical Report, October 2012.

[19] Nokia Siemens Networks, Nokia (2012) R1-124178, Need for signaling quasi co-location between CRS and CSI-RS. Nokia, Technical Report, October 2012.

[20] Alcatel Lucent, Alcatel Lucent Shanghai Bell (2012) R1-123155, Remaining aspects of quasi-co-located antenna ports. Alcatel, Technical Report, August 2012.

[21] 3GPP (2011) Coordinated multi-point operation for LTE physical layer aspects, 3GPP TR 36.819, v11.1.0. Third Generation Partnership Project, Technical Report.

[22] Scalia, L., Biermann, T., Choi, C.C., Kozu, K., and Kellerer, W. (2011) Power-efficient mobile backhaul design for CoMP support in future wireless access systems. In *Proceedings of IEEE INFOCOM*, April 2011.

[23] Chenyang, Y., Shengqian, H., Xueying, H., and Molisch, A.F. (2013) How do we design CoMP to achieve its promised potential? *IEEE Wireless Communications Magazine*, **20**(1), 67–74.

[24] Je, H., Lee, H., Kwak, K., Choi, S., Hong, Y.J., and Clerckx, B. (2011) Long-term channel information-based CoMP beamforming in LTE-Advanced systems. In *Proceedings of IEEE GLOBECOMM*, December 2011.

Part Four

Upcoming Technologies

10

Dense Small Cell Deployments

10.1 Introduction

As discussed in Chapter 1 mobile data traffic is expected to grow tremendously in the future, especially in indoor and outdoor hotspot areas. In order to provide for this huge growth in traffic, several solutions have been standardized in LTE Releases 10 and 11 including network densification via overlaid pico- and femtocells and advanced techniques such as CoMP and FeICIC. However, these existing technologies have to evolve to cater to future data demand. Accordingly, 3GPP held a workshop on the future requirements and candidate technologies necessary for the evolution of LTE in June 2012 [1]. A vast majority of companies showed interest in further densification of network via enhanced small cells to meet the future capacity requirements for diverse applications and traffic types. As a result, the very first discussions on LTE Release 12 have focused on developing viable candidate technologies for efficient operation of small cells. (A 'small cell' generally refers to a cell whose transmit power is lower than a macrocell, for example, pico- and femtocells are both small cells.) The unique deployment scenarios considered in Release 12 include dense small cell clusters overlaid within a macro coverage area, realistic assumptions for backhaul connectivity, and traffic asymmetry in the uplink and the downlink directions. This chapter describes the initial developments in LTE Release 12 regarding state-of-the-art technologies for small cell enhancements.

10.2 Evolution of Small Cells

Prior to Release 12, LTE Releases 10 and 11 had considered the co-channel heterogeneous deployments of small cells with the macro. These deployments were limited to isolated small cells deployed within a macro coverage area to support bursty traffic. The focus of Release 12 is on more evolved small cell deployment scenarios that include a larger number of overlaid small cells, possibly operating on different carriers than macrocells, realistic backhaul constraints, and emphasis on real-time traffic for evaluations. The key differences between prior LTE releases and Release 12 are summarized in Table 10.1.

Heterogeneous Networks in LTE-Advanced, First Edition. Joydeep Acharya, Long Gao and Sudhanshu Gaur.
© 2014 John Wiley & Sons, Ltd. Published 2014 by John Wiley & Sons, Ltd.
Companion Wesite: www.ltehetnet.com.

Table 10.1 Differences between Release 12 and previous LTE releases

	Previous releases	Release 12
Macro UE density	Low (~ 10)	High (~ 25)
Small cell clusters	Absent	Present
Dominant interference	Macro ↔ Pico	Macro ↔ Pico, Pico ↔ Pico
Backhaul connectivity	Ideal	Non-ideal
Traffic characteristics	Full buffer	Finite buffer
Dual connectivity	Not allowed	Allowed
New carrier type	Not available	3.5 GHz
TDD UL/DL configuration	Static	Dynamic

Figure 10.1 shows a conceptual small cell deployment scenario being considered in Release 12. The key aspects of small cell deployments are classified as described in the following [2].

- *Co-channel and separate frequency deployment of small cells*: Both co-channel and separate carrier deployment of small cells under the macro coverage area have been considered. As illustrated in Figure 10.1, the macrocell is deployed in legacy carrier F1. Small cells can be deployed in F1 which is the co-channel case or in a different carrier F2. One of the most important uses concerns the utilization of higher-frequency bands such as F2 = 3.5 GHz, unlikely to be deployed by the macrocells due to their smaller coverage. Such frequency bands are ideal for small cell deployments to boost network capacity.
- *Macro coverage*: Deployment scenarios are being considered where small cells can be deployed without concurrent coverage of a macrocell as shown in Figure 10.1, when a small cell is deployed in F2.
- *Small cell density*: In addition to sparse small cell deployments, dense deployments are being considered given their potential to offload more user traffic intelligently. As shown in Figure 10.1, network densification via the deployment of multiple small cell clusters covering a hotspot area is being considered. Dense small cell deployment is one of the key aspects of Release 12 and requires several enhancements at the physical layer such as interference coordination among the small cells in order to realize throughput gains.
- *Outdoor and indoor*: Prior LTE releases have considered both indoor and outdoor deployments of small cells. LTE Release 12 continues to focus on both these deployments.
- *Backhaul connectivity*: Practical constraints are unlikely to allow ideal backhaul connectivity between macrocells and small cells as well as between small cells. These backhaul limitations are likely to impact the choice of potential solutions for efficient operation of small cells. Release 12 small cell deployments will cover both ideal backhaul such as dedicated optical fiber connections as well as non-ideal backhaul with limited capacity and finite delay such as xDSL, microwave, etc.
- *Traffic characteristics*: Dense small cell deployments are expected to have a small number of associated UEs per small cell due to small coverage. This is likely to result in a highly fluctuating traffic in a small cell. In addition, traffic is expected to be highly asymmetrical in the uplink and downlink directions. Realistic buffer traffic to model real-time applications have higher priority in Release 12 evaluations.

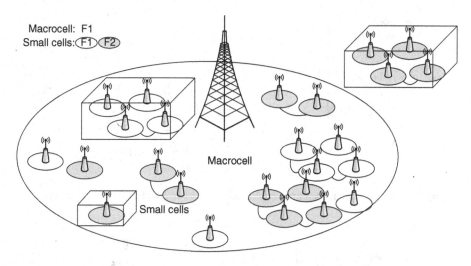

Figure 10.1 An example of target deployments for small cell enhancements considered in LTE Release 12

10.2.1 Deployment Scenarios

In order to validate the benefits of small cell enhancements, such as improvements in spectral efficiency and mobility enhancements, the following evolution path for small cell scenarios has been identified in Release 12 for evaluation [3].

Scenario 1 Small cells deployed in the same carrier (F1) as the macrocell.
- Clustered outdoor deployment of small cells with 4–10 small cells per cluster.
- Coordination between macro and small cells via both the ideal and non-ideal backhaul.
- Realistic buffer traffic models are prioritized.

Scenario 2 Small cells deployed in carrier (F2) different from the macrocell carrier (F1).
- Clustered deployment of small cells similar to Scenario 1. Both outdoor (Scenario 2a) and indoor (Scenario 2b) deployments have been considered.
- Non-ideal backhaul-based coordination is prioritized.
- Legacy UEs can access the small cell as a legacy cell.
- Realistic buffer traffic models are prioritized.

Scenario 3 Standalone deployment of small cells.
- Small cells deployed as a stand-alone cell regardless of macrocell coverage.
- Coordination among small cells via both the ideal and non-ideal backhaul.
- Realistic buffer traffic models are prioritized.

Figures 10.2, 10.3, and 10.4 illustrate example deployments for the scenarios described above. Note that since sparse co-channel scenarios have been extensively studied in the previous LTE releases, Release 12 focuses mostly on deployments with dense small cells operating on separate carriers to the macrocell. However, since dense co-channel scenarios have not been studied before, these are also included for evaluation. Table 10.2 shows the

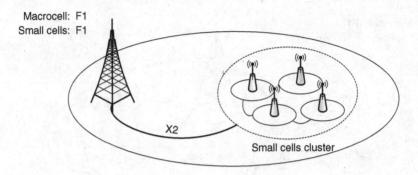

Figure 10.2 Co-channel deployment of macrocell with overlaid outdoor small cell clusters

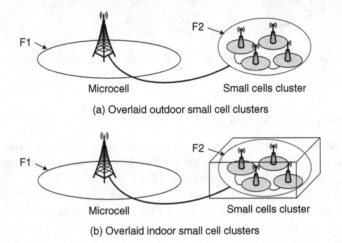

(a) Overlaid outdoor small cell clusters

(b) Overlaid indoor small cell clusters

Figure 10.3 Deployment of macrocell with overlaid small cell clusters on separate frequencies

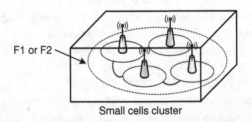

Figure 10.4 Deployment of standalone indoor small cell clusters with no coordination with macrocell

correspondence between small cell scenarios for evaluation in LTE Release 12 and real-life small cell deployments.

Figure 10.5 shows the macro association ratios, i.e. the fraction of total number of UEs that are associated with the macro for small cell scenario 2a where the macro and small cells are in different frequencies. The association ratio is evaluated for different association methods that

Table 10.2   Envisioned small cell deployment scenarios

Real-life deployment	Relevant small cell scenario
Train station	1, 2
Shopping mall	1, 2
Convention center	1, 2
Hotspots	2
Dense urban	1, 2
Sparse rural	3

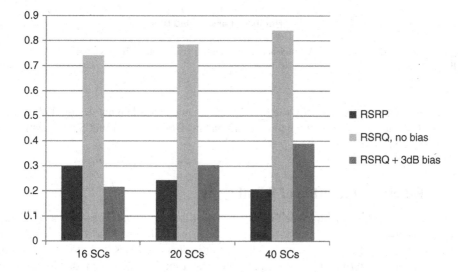

Figure 10.5   An example of macro association ratios in scenario 2a with 4 small cells/cluster

have been considered in Release 12. Traditional RSRP-based association methods yield a low association ratio for the macro as the small cells are clustered around locations of high UE density, thus leading to higher small cell RSRPs. However, this does not capture the information that the clustered nature of small cells also lead to higher interference and thus low SINR in the small cell frequency; this information is captured in RSRQ. For RSRQ-based associations, the macro association ratio is therefore highest. Increasing the number of small cell clusters increases the chances of higher small cell RSRP but also decreases the RSRQ further. This explains the trend as the number of clusters is varied. Other association methods such as RSRQ+bias yield intermediate association ratios compared to the extreme examples of RSRP and RSRQ.

An important feature of a dense small cell system is that there will be fewer UEs associated per small cell than traditional systems which have a lower number of small cells. For example, Figure 10.6 shows the distribution of small cells as per the number of UEs associated with them based on the RSRQ criterion [4]. The figure shows that the majority of small cells do not have any associated UEs. While this may seem reasonable, it has important consequences

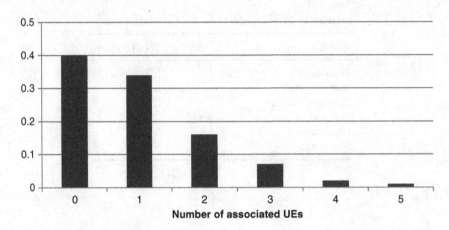

Figure 10.6   A distribution of the number of associated UEs/small cell in scenario 2a with 4 small cells/cluster

on the performance of small cells and also the associated control signal design. For example, is there any point in keeping a small cell that has no UEs switched on? Such questions are addressed in Section 10.3.

10.3   Efficient Operation of Small Cells

Network densification via small cell deployments offers opportunities as well as challenges that must be overcome to enhance network performance. On the one hand, dense and clustered deployment of small cells along with realistic traffic characteristics leads to a smaller number of active UEs offering opportunities for throughput improvement via reduction in control signaling overheads or use of higher order modulation (e.g. 256 QAM) [5, 6, 7].

On the other hand, dense small cell deployments also give rise to a number of issues as described in the following.

- *Interference among small cells*: Prior LTE releases have only considered interference problems for sparsely deployed small cells. The features developed, such as eICIC and FeICIC, protect small cells from the dominant interference arising from the downlink transmission of a macrocell. The interference conditions in clustered small cell deployments are more complex due to the fluctuating interference arising not only from the macrocell but also from strongly coupled small cells. Additionally, the use of dynamic UL/DL configurations in TDD leads to several new interference conditions that have not been studied in prior LTE releases (see Chapter 11 for a detailed discussion on dynamic TDD). Apart from the obvious impact on throughput, small cell interference will also have an adverse impact on the discovery of small cells as well as mobility management.
- *Impact on operation of legacy CoMP and CA mechanisms*: Limited coordination via non-ideal backhaul and network synchronization issues are likely to render tight coordination among cells infeasible. This will adversely impact CoMP and CA operations. For example, in case of CA an imperfect backhaul will necessitate the transmission of uplink

control signals (such as ACK/NACK and CSI information) on the SCell which will require changes in the existing Release 11 standard (see Section 12.3.1).

- *Impact on core network*: With dense small cell deployments, frequent and unnecessary handovers (between small cells) are expected to increase even for low UE mobility because of the smaller coverage area of small cells. This increase in handovers, and thus CN signaling load, will be directly proportional to the number of small cells in a cluster. Moreover, co-channel deployments of macro and small cells will result in severe interference conditions resulting in increased handover failure (HOF) rate compared to a homogeneous macrocell deployment, as shown in Figure 10.7 [3].

To address these issues, Release 12 has considered several potential technologies for efficient operation of small cells. Table 10.3 shows the relationship between some of the candidate technologies and their applicability to particular small cell scenarios. We discuss these solutions in detail in the remainder of this chapter.

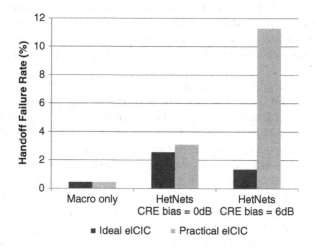

Figure 10.7 Handover failure (HOF) rates for macro-only and heterogeneous deployments

Table 10.3 Potential technologies for small cell enhancements in Release 12

Potential technologies	Scenario 1	Scenario 2a/2b	Scenario 3
Dual connectivity	√	√	×
Small cell discovery	×	√	×
Small cell ICIC	√	√	√
Overhead reduction	√	√	√
Higher modulation	√	√	√
NCT operation	√	√	×
Dynamic TDD operation	√	√	√

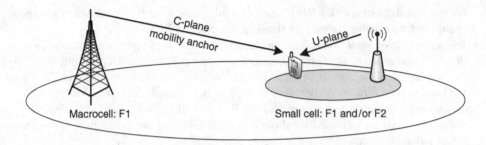

Figure 10.8 Dual connectivity via control-plane/user-plane split for efficient mobility management

10.3.1 Dual Connectivity

Dual connectivity is a key concept for small cell evolution in LTE. It implies that a UE has the ability to maintain simultaneous connections to both the macrocell and the small cell. A dual-connectivity-capable UE may at least perform RRM/RLM measurements at both the macro and small cell layers. Additionally, it may also receive downlink transmissions and possibly also perform uplink transmissions at macro and small cell layers, either simultaneously or in a TDM manner. The key benefits of dual connectivity are summarized in the following.

- *Mobility enhancement*: Dual connectivity via control-plane/user-plane split between the macro and small cell layers is expected to help achieve efficient mobility management. In this case, the UE receives the control-plane transmission and user-plane information from different nodes, distinguishing this technology from, say, CoMP (where there is no separation of the two planes). Figure 10.8 shows an example usage of dual connectivity that allows a UE to maintain its RRC connection with a macrocell while it receives uplink/downlink data from the small cells. The advantage of this technology is that as the UE moves between coverage areas of the small cells, it can receive data from the nearest small cell at any given instant. However, since it receives control-plane data from the macro there is no handover when it switches between small cells. New signaling mechanisms have to be defined to enable this, but they can be designed to incur less delay and overheads than traditional handover.
- *Throughput enhancements*: Dual connectivity can allow an advanced UE to receive PDSCH transmissions from macro and small cell layers simultaneously, thereby improving throughput performance.
- *UL/DL power imbalance*: Small cell deployments with overlapping macro coverage are characterized by UL/DL power imbalance for UEs connected to a small cell. For example, it is likely that a UE which is closer to a small cell will have a stronger UL connection with the small cell node but a stronger DL connection with the macro eNodeB due to higher transmission power at macro. For heavily loaded macrocells, it may be beneficial to offload UL traffic to the small cell layer while keeping the DL traffic in the macro layer. Such offloading is beneficial for Scenarios 1 and 2 and can be realized via dual connectivity as illustrated in Figure 10.9.

Dual connectivity has different levels of specification impacts depending on whether the inter-eNodeB connectivity is for a co-channel deployment or for deployments in different

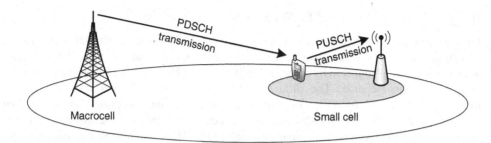

Figure 10.9 Dual connectivity for addressing UL/DL imbalance

frequency bands. For instance, for inter-band deployments (Scenario 2) the expected benefits of dual connectivity will largely depend upon the UE capability of simultaneous reception and/or transmission to both the macro and the small cell. For co-channel deployments (Scenario 1), dual connectivity may be limited to enabling UL/DL decoupled operation. Consequently, dual connectivity operation will have different requirements for the physical layer enhancements. Based on the UE capability, three modes of dual connectivity are envisioned in Release 12. These are captured in Table 10.4 and are explained in detail as follows.

1. **Dual Connectivity Mode 0**: This mode supports mobility enhancements only and does not involve multiple data connections at the macro and small cell layers. Instead, the UE maintains data connections on the small cell layer and is configured to perform mobility-related measurements on the macrocell layer. Such an architecture is also popularly known as control/user (C/U) split.
2. **Dual Connectivity Mode 1**: In addition to mobility enhancements, this mode also supports inter-eNodeB downlink-only carrier aggregation (CA) and allows a UE with appropriate CA capability to aggregate multiple downlink transmissions from both macro and small cell layers.
3. **Dual Connectivity Mode 2**: This mode supports inter-eNodeB uplink/downlink carrier aggregation. A UE can therefore aggregate multiple downlink transmissions from both macro and small cell layers as well as perform uplink transmission in both the layers.

Note that in addition to the UE capability assumption, dual connectivity gain also depends on the degree of network synchronization accuracy.

Table 10.4 UE capability requirements for supporting different modes of dual connectivity

Dual connectivity mode	UE capability requirement
Mobility enhancements	User-plane operation on small cell layer, control-plane measurements on macro layer.
Inter-site DL-only CA	Simultaneous DL reception on both layers, TDM-based UL transmission to both layers.
Inter-site UL/DL CA	Simultaneous DL reception and UL transmission to both layers.

10.3.1.1 Impact of Non-Ideal Backhaul on Dual Connectivity

The carrier aggregation mechanism of Release 10 and CoMP feature introduced in Release 11 can be categorized under dual connectivity. While CA Scenario 4 is an *intra-eNodeB inter-frequency* dual connectivity scheme, CoMP Scenario 4 is an *intra-eNode intra-frequency* dual connectivity scheme. The difference between Release 12 dual connectivity and Release 10/11 dual connectivity lies in the required backhaul support. Release 12 dual connectivity assumes finite backhaul delay between cells which can be as large as 60 ms [2], rendering tight coordination among cells infeasible. Thus, a non-ideal backhaul requires macro and small cells to perform independent resource scheduling along with independent feedback signaling (such as ACK/NACK, CSI transmission) [8–10].

It is expected that Release 12 will specify support for Uplink Control Information (UCI) transmission over PUCCH in each SCell as shown in Figure 10.10. This will be a key differentiating factor between the legacy CA and inter-site CA-based dual connectivity. Note that since macro and small cells can have different coverage, PUCCH-based uplink power control may need to be modified to support independent power control processes. For dual connectivity Mode 1, time-switching of uplink transmissions between the macro and small cells will need to be supported, along with tight synchronization between the serving eNodeBs.

10.3.2 ICIC Mechanism

Increasing the density of small cells is an effective mechanism to significantly improve system throughput. However, deployment of a large number of small cell nodes will increase the power consumption not only at the eNodeBs but also for the UEs; they will need to discover a large number of small cells and therefore spend more time performing RRM/RLM measurements instead of transitioning into sleep mode. Also, given the low number of active UEs per small cell, some of these small cell nodes are unlikely to serve any UEs at certain time periods but continually transmit CRS and other control channels, causing considerable interference at the neighboring cells. There are two possible sources of interference, namely: interference between a macrocell and a small cell (Scenario 1), and interference among small cells (Scenarios 1–3). Such small cell interference scenarios can be differentiated from those considered in Release 10/11 deployments as follows.

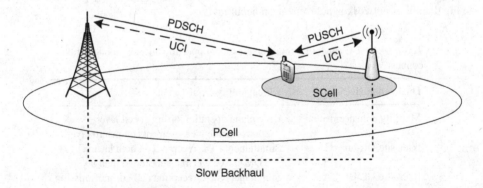

Figure 10.10   Dual connectivity with non-ideal backhaul

- In Release 10/11 HetNet deployments, the interference at a UE served by a small cell mainly comes from the downlink transmission in the macrocell. The interference contribution from other picocells is usually neglected due to sparse deployment of picocells. In dense deployment of small cells, interference from macrocells is no longer the single dominant interference source. Rather, the interference between small cells can be significant depending upon the number of small cells per cluster as well as the number of small cell clusters in a macro area. Legacy interference mitigation mechanisms, such as muting of the macro or transmission of ABS subframes, will therefore not be sufficient.
- Small cells within a cluster will observe severe interference in various control channels/signals such as PSS/SSS, PBCH, and PDCCH due to co-channel transmissions from within the small cell clusters. Legacy ICIC solutions are only applicable for data channels, and will not alleviate such interference issues affecting control channels.
- Given the highly dynamic traffic in small cell deployments, interference from a given source (macro or small cell) may fluctuate considerably in time. This may lead to non-accurate CSI feedback, and also adversely impact receiver processing such MMSE that are based on interference averaging.
- Release 12 has been considering more diversified backhaul connectivity between macrocells and small cells as well as between small cells. The benefits from interference coordination schemes such as eICIC/FeICIC or CoMP, which require ideal backhaul connectivity, may therefore be limited or difficult to achieve.

A relatively simple mechanism to avoid the above-mentioned interference issues is to reduce or suspend the transmission of CRS and control channels when there are no active UEs in the small cells. Recall that CRS REs are power boosted for reliable channel measurements, and cause the most severe inter cell interference to PDSCH transmissions in small cell clusters. One straightforward way to reduce CRS interference is by utilizing legacy mechanism of configuring subframes as MBSFN which carry reduced CRS. However, the number of subframes that can be configured as MBSFN subframes are limited. Further, CRS transmission cannot be completely eliminated in MBSFN subframes. Another issue is that the number of MBSFN subframes cannot be changed rapidly which could result in some restrictions in case the number of legacy UEs requiring a CRS transmission mode increases. The use of MBSFN mechanism therefore has limited benefits. As a result, Release 12 has been considering new potential solutions for the mitigation of inter-cell interference among small cells. In the following sections, we discuss some of the candidate interference management technologies.

10.3.2.1 Small Cell On/Off

Small cell On/Off is one of the key candidate technologies for reducing co-channel interference between small cells [11, 12, 13]. As the name suggests, it requires small cells to maintain two distinct states, *On* and *Off*, as shown in Figure 10.11. During the On state the small cell transmits the CRS along with other legacy control signals to facilitate RRM/RLM measurements and data transmission. During the Off state, the cell either does not transmit legacy control signals or transmits a reduced set of legacy control signals. In this state, a UE cannot access the cell for any data transmission or RRM/RLM measurements. For lightly loaded small cells, it may be desirable to hand over their UEs to the neighboring cells and then switch them to Off state. As a result of switching off, the overall interference reduces and the network throughput

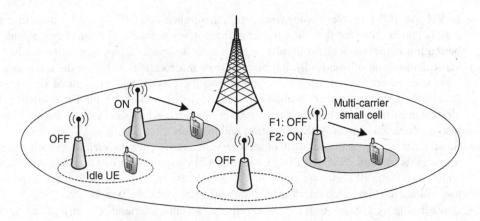

Figure 10.11   Inter-cell interference management via small cell On/Off mechanism

performance can improve for light or medium traffic load. A small cell in Off state may be switched On to support the neighboring small cells if their traffic load becomes high.

The criteria for transitioning a small cell from On to Off and vice versa are implementation specific. For example, transition to Off may be based on the total uplink/downlink data buffer at the eNodeB if it is low. It may be based on interference reports from the neighboring small cell nodes which states that the small cell is creating too much interference. Transition from the Off to On state may be triggered if the traffic loads of the neighboring cells exceed some pre-set threshold. For a small cell enabled with multiple carriers, On/Off mechanism could be applied to a subset of carrier frequencies (instead of switching off the small cell completely) by utilizing the legacy OCS mechanism (see Chapter 7). These carrier frequencies act as capacity enhancement layers in the network and can be activated or deactivated based on the prevailing capacity needs of the network.

The achievable throughput gain will depend on how dynamically cells can be switched On/Off. If a small cell can switch off its transmission as soon as there are no UEs to be served and switch back on instantly when one or more UEs enter its coverage area, system throughput is maximized. As an example, Figure 10.12 shows that dynamic small cell On/Off (i.e. subframe-level On/Off switching) can significantly boost the network throughput under varying load conditions and varying fraction of MBSFN subframes [12]. However, subframe-level On/Off may not be possible due to practical constraints. For example, there is a large latency involved from the time a network determines that an Off cell should be switched On to the time when that cell resumes data transmission. The latency can be of the same order of magnitude (or even larger) as the duration required for the transmission/reception of a data packet [14]. Similarly, the latency between the time a network determines that a cell should be switched Off to the time when it completely switches Off can run from several milliseconds to a few hundreds of milliseconds depending upon the number of UEs that need to be handed over to other cells. As a result, a practical network will adopt a conservative approach to switch Off a cell which will reduce the underlying throughput gains of dynamic On/Off mechanism. Table 10.5 shows the feasible timescales for small cell On/Off based on current LTE specifications.

Note that the On/Off mechanism will have an impact on the mobility performance and core network signaling. For UEs not capable of dual connectivity, handover procedure will be

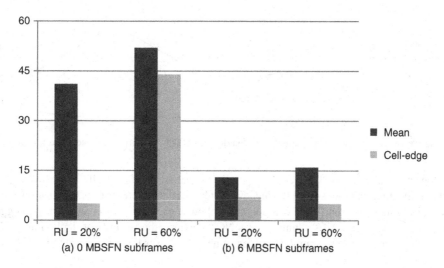

Figure 10.12 Percentage gains with dynamic On/Off scheme applied to Scenario 1 with 4 small cells per cluster for a different number of MBSFN subframes and resource utilization

Table 10.5 Feasible timescales for small cell On/Off based on current LTE specifications

Event	Explanation	Non-CA-capable UE	CA-capable UE (small cell configured as SCell)
1	Time before a just turned On small cell can be used	2000–4000	500–1000
2	Time before an already On small cell can be used	100–150	80–120
3	Time required in switching Off a cell	100–150	100–150

needed before they can access a just-turned-On small cell whereas dual-connectivity-capable UEs can avoid handover in such cases.

10.3.2.2 New Carrier Type

Almost complete CRS interference avoidance can be achieved by the New Carrier Type (NCT) operation which was studied in 3GPP as part of Release 12 feature. An eNodeB deploying a NCT carrier (e.g. 3.5 GHz) does not need to transmit any control signal such as CRS in most of the subframes when there is no traffic in the cell. The legacy CRS transmission is replaced by much more sparse CRS transmission that occurs once every 5 ms. NCT can therefore potentially achieve similar reduction in CRS interference as small cell On/Off without actually having to switch off the small cell. However, note that NCT does not allow the legacy UEs to access the cell and also involves significant modifications to specifications.

NCT can operate in two modes: *non-standalone* and *standalone*. In the former, it is always operated along with a legacy carrier. For example, for a UE capable of carrier aggregation,

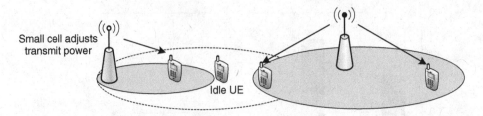

Figure 10.13 Interference management via downlink TPC to dynamically adjust cell coverage

the legacy carrier can be the PCell and the NCT can act as the SCell. This way important control information comes from the legacy carrier and the NCT carrier is used for throughput maximization. Standalone NCT can operate without legacy assistance, as the name suggests. However, since legacy UEs cannot access NCT carriers, the benefits of deploying NCT are directly proportional to the number of UEs of newer releases that can operate on NCT [15].

10.3.2.3 Transmit Power Control

One possible approach for managing interference within a small cell cluster is by utilizing downlink transmission power control (TPC) to change cell coverage dynamically as per the prevailing traffic load within the small cell cluster [16]. This is shown in Figure 10.13. The power control applied at an eNodeB can be based on the interference information reported by the victim UEs of other eNodeBs or measured directly by the eNodeBs by measuring the uplink signals from the victim UEs, such as SRS. For efficient and accurate power control, small cells within a cluster will need to co-ordinate among themselves to decide their transmit power to mitigate the inter-cell interference while ensuring best possible coverage, capacity and mobility performance for their own UEs. Note that dynamic cell On/Off mechanism is an extreme case of downlink TPC.

10.3.2.4 Enhancement of Legacy ABS Mechanism

Recall that Release 11 FeICIC technique protects picocells from the downlink interference of a macrocell by configuring ABS subframes on macrocells. It is usually used in conjunction with cell range expansion (CRE) for offloading traffic to the small cells. However, for dense small cell deployments macrocells are not the major interference sources. Interference among small cells can be significant. Therefore, one straightforward enhancement of FeICIC technique is to allow interfering small cells to use ABS configuration on certain subframes [17, 18]. Note that given the highly dynamic traffic pattern in small cells, the ABS patterns will need to be updated frequently. Small cells therefore need to coordinate over the backhaul to adapt their ABS patterns. However, a slow backhaul will limit the performance gains of such FeICIC schemes.

10.3.3 Small Cell Discovery

As discussed in previous sections, Release 12 small cell deployments will be characterized by clustered small cells with significantly overlapping coverage areas as well as a high level of

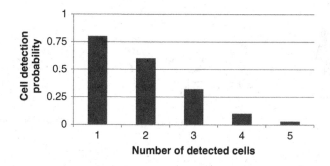

Figure 10.14   Cell detection probability for Scenario 2a with 4 small cells per cluster

synchronization among small cells within the cluster. A UE in a small cell cluster coverage is therefore likely to be subjected to radically different interference conditions in the discovery signals (involving PSS/SSS collisions) when compared to the conventional homogeneous and heterogeneous networks. As an example, Figure 10.14 shows the impact of PSS/SSS collisions for small cell Scenario 2a [19]. The figures shows that even although there are 4 small cells in a cluster, very few small cells are actually detected by a UE due to interference. Apart from inter-cell interference issues, small cell discovery mechanisms also need to address the issues described in the following.

- *UE power consumption*: One of the key methods of enhancing network throughput in small cell deployments is to offload the UEs to the small cells in a timely manner once they move into the small cell coverage region. This requires a UE to spend considerable battery resources performing regular measurements to detect a large number of small cells. This is more acute when there are multiple carriers and the small cells are deployed on another carrier than the current UE association. This can have an adverse impact on UE power consumption, especially for the UEs which are outside the small cell coverage but still perform measurements that are ultimately not necessary.
- *Detection of cells in Off state*: As discussed in the previous section, small cell On/Off mechanism is a potential ICIC mechanism for small cell deployments. However, the transition of a small cell to Off state will make its timely discovery by newly arrived UEs or the UEs in the IDLE mode challenging. In particular, the discovery of small cells that go into Off state for a long time is even more challenging.
- *Increased PCI collision/confusion*: Another issue that arises from small cell densification is the limited number of PCIs available for assignment in a LTE network. The number of PSS are limited to 3 while SSS sequences are 168 in number, resulting in a total of 504 supported PCIs. A dense small cell deployment is therefore likely to have cases when two or more small cells in the same cluster are assigned the same PSS and SSS sequences (i.e. PCI) resulting in PCI collision. Note that even if the assigned SSS sequences are different, PSS collision may also greatly impact the coherent detection of SSS. These issues will adversely affect the ability of a UE to discover small cells in a timely manner. Another related issue is PCI confusion, as shown in Figure 10.15, where the network cannot uniquely distinguish the cell detected by the UE due to the same PCI used in multiple cells [20].

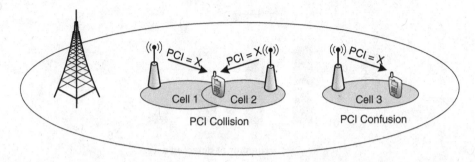

Figure 10.15   An example of (a) PCI collision where synchronization signals collide at UE, and (b) PCI confusion as the network fails to determine which cell UE connects to

The SINR requirements specified in current LTE standards for cell detection are quite stringent as they focus on the worst-case scenarios. For example, a cell is considered detectable if the received power of PSS/SSS and CRS exceeds −6 dB threshold for co-channel deployments and −4 dB for inter-frequency deployments. One straightforward way to improve the legacy mechanism is to relax the stringent SINR requirements for cell detection in small cell scenarios. This would enable a UE to detect a cell with fewer samples of the received PSS/SSS sequences, therefore taking a shorter time for discovery. For the efficient discovery of small cells using On/Off mechanism, the following potential solutions are being considered [21, 22].

- *Discovery signal*: A small cell in Off state may transmit a discovery signal with a long DTX cycle (e.g. sent once every 100 ms) to inform UEs of its presence. Upon detecting the signal, the UE could either try to wake the small cell or use the macro's assistance to determine its cell ID and the specific time at which the dormant cell will wake up and accept connection requests. The discovery signal can be formed from legacy signals, such as PSS/SSS, CRS, and PRS, or it can be newly designed. The DTX cycle length of the discovery signal should take into account the trade-off between the offloading potential and the energy efficiency. Corresponding UE measurement procedures will need to be specified depending on the discovery signal design. Additionally, in order to enable UEs to detect a large number of small cells within one measurement period, the small cells will need to transmit the discovery signals synchronously.
- *Uplink channel monitoring*: Another mechanism for improving small cell discovery involves uplink channel monitoring by the cells in the Off state. This assumes that cells in Off states can still periodically listen to uplink transmissions from UEs that are associated with the neighboring cells. The wake-up operation is based on the dormant eNodeB detecting the uplink transmissions (e.g. PRACH, SRS, etc.) of these UEs and reporting details of the detected signaling back to the macro or a nearby active small cell node. The recipient cell can then reactivate the neighboring small cell nodes that are in the Off state as well as assist them to identify the UEs. Since UL-based solutions require the dormant cell coverage area to be fully or partially under the macro (or small cell) area coverage, they cannot be effective in all cases. Furthermore, the UL-based solutions will require a small cell node in the Off state to monitor uplink transmissions at regular intervals, thereby compromising the benefits of energy saving.

10.4   Control Signaling Enhancement

The amount of downlink control information (DCI) transmitted via PDCCH or EPDCCH depends on the number of active UEs of the cell requiring uplink and downlink scheduling. Given that densely deployed small cells are most likely to have 1 or 2 associated UEs per small cell (as seen in Figure 10.6), only a limited amount of control signaling is required in small cells compared to the macrocell which typically has a much larger number of UEs associated with it [23–25]. If PDCCH is used in such scenarios, the resulting control signal region will comprise at least 1 OFDM symbol (PDCCH resource granularity) resulting in ≈7.14% overheads irrespective of the system bandwidth. Legacy PDCCH is therefore not efficient for DCI transmission in small cells if overheads are sufficient.

On the other hand, EPDCCH-based control signaling can use less overheads compared to PDCCH. For instance, EPDCCH transmission has a granularity of 1 PRB pair that corresponds to overheads of 1% and 4% for system bandwidths of 5 and 20 MHz. A possible method of reducing downlink control overheads is by removing the legacy control region (PDCCH) and relying on EPDCCH for the transmission of DCI in small cells. Note that Release 12 already supports control region overhead reduction in NCT deployments which do not carry any PDCCH. However, unlike NCT, the downlink control region cannot be entirely removed from all subframes for legacy carriers; otherwise, it would render the network inaccessible to the legacy UEs. Instead, Release 12 studies have considered utilizing PDCCH in legacy subframes to schedule control information for other subframes from which legacy control regions have been removed by means of multi-subframe scheduling (one control channel schedules UL/DL transmissions over multiple subframes) or cross-subframe scheduling (one subframe contains multiple control channels scheduling UL/DL transmissions in multiple subframes). During Release 12 studies, different companies provided diverse views on the benefits of these schemes [26, 27, 28, 29, 30]. In the following sections we discuss these two schemes along with their pros and cons.

10.4.1   Multi-Subframe Scheduling

A small cell is likely to schedule the same set of UEs in consecutive subframes due to the time-invariant channel conditions and smaller number of associated UEs. As a result, various scheduling parameters carried by DCI, such as allocated RBs, MCS, and PMI, may not vary by much across consecutive subframes. For such case, multi-subframe scheduling can effectively reduce the control channel overhead by limiting PDCCH transmission to the 1st subframe of the multi-subframe resource allocation and using it to schedule UL/DL transmissions for all the remaining subframes.

Figure 10.16 depicts an example of multi-subframe scheduling where PDCCH is present only in 1 out of 4 subframes of the multi-subframe resource allocation, thus significantly reducing control channel overheads. Additionally, multi-subframe scheduling can improve channel estimation performance by collecting the received signals over multiple subframes. This is especially useful for compensating degradation in channel estimation performance due to reduced-density DMRS per subframe. These benefits make multi-subframe scheduling a promising technique for improving spectral efficiency of small cell deployments. Multi-subframe scheduling has some similarities with legacy semi-persistent scheduling

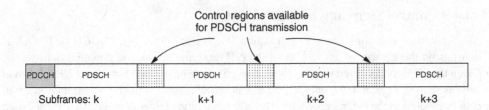

Figure 10.16   Multi-subframe scheduling

Table 10.6   Differences between multi-subframe scheduling and semi-persistent scheduling

	Semi-persistent scheduling (SPS)	Multi-subframe scheduling
Supported applications	Periodic traffic with small packet size (eg. VoIP)	Any application with large packet sizes (e.g. best effort)
Suitable channel	All types of channel	Slow time varying
Channel estimation	Not useful as scheduled subframes are separated greatly in time	Enhances channel estimation for reduced DMRS due to contiguous subframe allocation
Link adaptation	Not supported	Supported
Control signaling	One DCI each to activate and deactivate SPS	One DCI to indicate resource grant

(SPS) as both provide scheduling of multiple subframes. However, there are also significant differences between the two mechanisms as elaborated in Table 10.6.

Despite these benefits, multi-subframe scheduling adversely impacts the scheduling flexibility of an eNodeB [27, 28, 29]. For instance, an eNodeB serving a multi-subframe-configured UE loses the ability to handle bursty traffic situations that may arise in the middle of multi-subframe allocation. It also prevents the eNodeB from making appropriate scheduling decisions when interference situations change due to small cell On/Off. Similarly, the ability of the eNodeB to perform load balancing across multiple available carriers is also impacted by multi-subframe scheduling. Besides scheduling issues, multi-subframe scheduling may also require some enhancements in legacy DCI formats to support additional HARQ indices, RV, and NDI fields for proper HARQ operation across the subframes for which the multi-subframe grant is applicable [30]. This may potentially increase the complexity for blind decoding at the UE. Also, note that the throughput benefits offered by multi-subframe scheduling are maximized when only the 1st subframe of the multi-subframe allocation carries the control region and the the PDSCH region of the remaining subframes starts from the 1st OFDM symbol. This is possible to achieve for NCT but requires specification change for legacy carriers.

10.4.2   Cross-Subframe Scheduling

In cross-subframe scheduling the control region in one subframe carries DCIs for multiple subsequent subframes, thereby eliminating the need for control region in those subframes [26, 27]. This enables the control channel scheduler to optimize the resource usage by reshaping the control region such that fragmentation of PDCCH/EPCCH resources is minimized. For example, for DCI transmission via PDCCH, consider a scenario where a small cell needs to schedule only a couple of UEs per subframe. In such cases, the PDCCH region may remain under-utilized, especially for a larger system bandwidth of, say, 20 MHz. Cross-scheduling can avoid this wastage of resources by allowing DCI transmission corresponding to multiple consecutive subframes in a single subframe. Similar benefits can be observed for EPDCCH-based DCI transmission.

Note that both cross-subframe scheduling and multi-subframe scheduling can be supported together to efficiently managing the number of DCIs and their resource mapping. Unlike multi-subframe scheduling, the HARQ aspects related to cross-subframe scheduling are the same as for legacy single-subframe scheduling.

10.5 Reference Signal Overhead Reduction

UE-specific reference signals (RS) in earlier LTE Releases were designed primarily for homogeneous macro deployments and provide good performance under various channel conditions and for low and high UE mobility. For small cell deployments, the associated UEs may be assumed to have low mobility either because they are indoors or because high mobility UEs can be offloaded to the macrocell. Limited UE mobility, coupled with small delay spreads in indoor environments, imply that the use of legacy UE-specific RS may not be optimal. For instance, the relatively flat channel in a small cell can be estimated reliably using the lower density of RS when compared to legacy RS. Any loss in channel estimation performance due to the lower density of RS can be compensated for by a high operating SNR of UEs in the small cell coverage. In addition to the potential benefits due to less overheads, transmission of reduced-density RS will cause less interference at the UEs scheduled in MU-MIMO transmission mode and the neighboring cells' UEs that are scheduled during the same time-frequency resources. The reduction of UE-specific RS density for small cell deployments has therefore been considered in Release 12. We discuss the specifics of uplink and downlink UE-specific RS density reduction in the following sections.

10.5.1 Downlink DMRS

For PDSCH transmissions based on TM7−10, DMRS is present in all the PRBs assigned to the UE to ensure reliable estimation of its downlink channel. The DMRS overhead depends on subframe type, CP type, and the rank of the corresponding PDSCH transmission. For PDSCH transmissions in normal subframes with normal CP there are 12 DMRS REs per PRB pair for rank 1−2 transmissions and 24 DMRS REs per PRB pair for rank 3 and higher transmission. This results in $\approx 7.1\%$ and $\approx 14.3\%$ overheads, respectively. In order to reduce this overhead, DMRS density reduction both in frequency and time domain can be considered. In particular, the following schemes (or a combination of them) can offer DMRS density reduction [31, 32, 33]:

- *Scheme 1*: Reducing DMRS density in each PRB. Figure 10.17 shows an example of reduced-density DMRS and also shows the legacy DMRS for comparison.

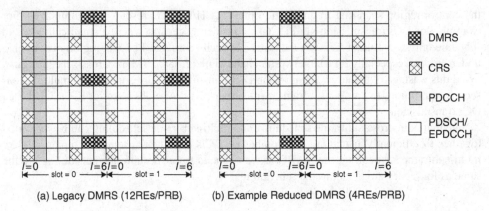

(a) Legacy DMRS (12REs/PRB)   (b) Example Reduced DMRS (4REs/PRB)

Figure 10.17   Legacy and reduced-density downlink DMRS patterns for rank up to 2

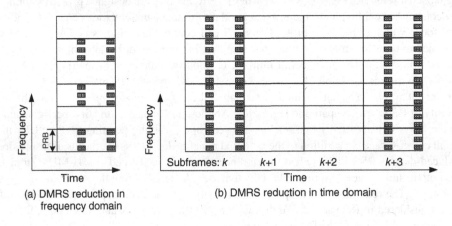

(a) DMRS reduction in   (b) DMRS reduction in time domain
frequency domain

Figure 10.18   Downlink DMRS density reduction along frequency and time domains

- *Scheme 2*: Reducing the set of PRBs within a subframe that carry DMRS as shown in Figure 10.18a.
- *Scheme 3*: If the UE has been allocated multiple consecutive subframes by multi-subframe scheduling as discussed in Section 10.4.1, then restricting DMRS to a subset of subframes among the allocated subframes assigned to a UE as shown in Figure 10.18b.

Note that the legacy DMRS sequence design ensures orthogonality which might be impaired by DMRS length reduction required by Schemes 1 and 2. To minimize this loss in orthogonality, one possible solution is to puncture legacy DMRS patterns to obtain reduced-density DMRS. Also, the variation of DMRS density within a subframe (Scheme 2) or across subsequent subframes (Scheme 3) may lead to fluctuating interference conditions for the UEs in adjacent cells. Evaluations performed during Release 12 have shown that the gain from reduced-density DMRS mainly occurs at high-SNR regime and is dependent upon PRB bundling size, UE speed, and transmission scheme.

10.5.2   *Uplink DMRS*

Similar to reduced-density downlink DMRS, overhead reduction for uplink DMRS can also benefit small cell deployments [34, 35]. Existing LTE releases specify uplink DMRS transmission in 2 OFDM symbols in each subframe as shown in Figure 10.19. For PUSCH transmissions in a subframe with normal CP, this results in DMRS overhead of $\approx14.3\%$ or $\approx15.4\%$ depending on whether SRS is scheduled in the last OFDM symbol. Similar to downlink DMRS reduction, a simple method is to consider using a frequency comb to reduce uplink DMRS overhead as shown in Figure 10.19. Another solution is to reduce DMRS density across subsequent subframes and utilize multi-subframe scheduling to improve channel estimation. This is illustrated in Figure 10.20 where the second DMRS is eliminated from each subframe, resulting in $\approx7.1\%$ overhead reduction. Note however that transmitting 1 DMRS symbol per subframe adversely affects DMRS multiplexing capacity as an orthogonal cover code (OCC) cannot be used. One way to recapture the lost capacity would be to apply

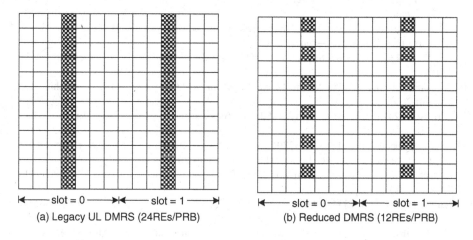

Figure 10.19   Legacy and reduced uplink DMRS in frequency domain

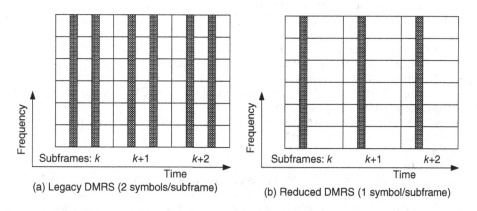

Figure 10.20   Legacy and reduced uplink DMRS in time domain

OCC across two consecutive subframes with reduced DMRS. As in the downlink scenario, issues such as fluctuating interference conditions and collisions between data and reference signals also arise in the case of uplink DMRS reduction.

References

[1] 3GPP (2012) RWS-120045, Summary of 3GPP TSG-RAN workshop on Release 12 and onward. Third Generation Partnership Project, Technical Report, June 2012.

[2] 3GPP (2013) TR 36.932, Scenarios and requirements for small cell enhancements for E-UTRA and E-UTRAN (Release 12). Third Generation Partnership Project, Technical Report, March 2013.

[3] 3GPP (2013) TR 36.872, Small cell enhancements for E-UTRA and E-UTRAN - physical layer aspects (Release 12). Third Generation Partnership Project, Technical Report, August 2013.

[4] Hitachi (2013) R1-131151, Initial Evaluation results on Small Cell Scenarios. Hitachi, Technical Report, April 2013.

[5] Hitachi (2013) R1-131146, Discussion on 256QAM in Small Cell Scenarios. Hitachi, Technical Report, January 2013.

[6] Huawei, HiSilicon (2013) R1-130893, Evaluation and Analysis for Higher Order Modulation. Huawei, Technical Report, January 2013.

[7] CATT (2013) R1-130987, Support of 256QAM for Small Cells. CATT Technical Report, January 2013.

[8] Hitachi (2013) R1-133558, Physical Layer Impact of Dual Connectivity. Hitachi, Technical Report, August 2013.

[9] Ericsson (2013) R1-133436, Physical Layer Aspects of Dual Connectivit. Ericsson, Technical Report, August 2013.

[10] Huawei, HiSilicon (2013) R1-132894, Physical Layer Support of Upper Layer Aspects of Small Cell Enhancements. Huawei, Technical Report, August 2013.

[11] Hitachi (2013) R1-133829, Evaluation of Small Cell ON/OFF Schemes. Hitachi, Technical Report, August 2013.

[12] Ericsson (2013) R1-133431, On Dynamic Small Cell ON/OFF. Ericsson, Technical Report, August 2013.

[13] LG Electronics (2013) R1-133769, Simulation Results on Small Cell ON/OFF Schemes. LG Electronics, Technical Report, August 2013.

[14] Panasonic (2013) R1-133790, Discussion on the Time Scales for Evaluations of Small Cell ON/OFF. Panasonic, Technical Report, August 2013.

[15] Ericsson (2012) Rp-122028, New Carrier Type for LTE. Ericsson, Technical Report, December 2012.

[16] NTT DoCoMo (2013) R1-132680, Performance Evaluation of ICIC with DL Transmission Power Adaptation for SCE. NTT DoCoMo, Technical Report, May 2013.

[17] ZTE (2013) R1-133073, ABS Enhancement for Small Cell. ZTE, Technical Report, August 2013.

[18] Samsung (2013) R1-133108, ABS Enhancements for Interference Mitigation in Small Cells. Samsung, Technical Report, August 2013.

[19] Intel (2013) R1-130920, Initial system level evaluation on small cell discovery using PSS/SSS/CRS. Intel, Technical Report, April 2013.

[20] Intel (2013) R1-130919, Discussion on small cell discovery operation. Intel, Technical Report, April 2013.

[21] LG Electronics (2013) R1-133377, Evaluation Results of Small Cell Discovery. LG Electronics, Technical Report, August 2013.

[22] Hitachi (2013) R1-130342, Efficient Small Cell Operation. Hitachi, Technical Report, January 2013.

[23] Samsung (2013) R1-132639, Control Signaling Enhancements for Small Cells. Samsung, Technical Report, May 2013.

[24] QualComm (2013) R1-132491, Control Channel Overhead Reduction. QualComm, Technical Report, May 2013.

[25] HTC (2013) R1-132077, Control Signalling Enhancements for Improved Spectral Efficiency. HTC, Technical Report, May 2013.

[26] Alcatel-Lucent Shanghai Bell, Alcatel-Lucent (2013) R1-132057, Consideration for Multi-SF and Cross-SF Scheduling in LTE-A. Alcatel, Technical Report, May 2013.

[27] Ericsson (2013) R1-132145, On Multi-Subframe and Cross-Subframe Scheduling. Ericsson, Technical Report, May 2013.

[28] China Telecom (2013) R1-132202, Discussion on multi-subframe scheduling. China Telecom, Technical Report, May 2013.

[29] Samsung (2013) R1-131023, Considerations on Multi-Subframe Scheduling. Samsung, Technical Report, April 2013.

[30] Nokia, Nokia Siemens Networks (2013) R1-132301, HARQ considerations in multi-subframe scheduling. Nokia, Technical Report, May 2013.

[31] Huawei, HiSilicon (2013) R1-130891, Discussion and Evaluation for DMRS Overhead Reduction. Huawei, Technical Report, April 2013.

[32] LG Electronics (2013) R1-131298, DM-RS Configurability for Small Cell. LG Electronics, Technical Report, April 2013.

[33] Hitachi (2013) R1-131147, DMRS Overhead Reduction for Small Cell. Hitachi, Technical Report, April 2013.

[34] ZTE (2013) R1-131052, Evaluation on the Uplink DMRS Overhead Reduction of Small Cells. ZTE, Technical Report, April 2013.

[35] Ericsson (2013) R1-131615, Evaluation on UL DMRS Overhead Reduction. Ericsson, Technical Report, April 2013.

11

TD-LTE Enhancements for Small Cells

One of the most important characteristics of small cell deployments is the highly dynamic nature of traffic coupled with asymmetry between uplink and downlink load. This chapter covers ongoing discussions in LTE Release 12 about TD-LTE enhancements for traffic adaptation in small cells. These enhancements include dynamic reconfiguration of TDD UL/DL subframe ratios in small cells, the feasible timescales and the signaling mechanisms required for TDD UL/DL reconfiguration, and interference mitigation schemes needed to counter new inter-cell interference conditions induced by dynamic TDD reconfiguration. In addition, this chapter also provides an introduction to the interworking mechanisms between FDD and TDD networks that allow a UE to connect to multiple frequency bands with different duplex modes simultaneously in order to further enhance its throughput performance.

11.1 Enhancements for Dynamic TDD

The TDD duplexing mode is an attractive alternative to the FDD duplexing mode for dense small cell deployments as it offers much higher flexibility to handle asymmetric traffic in the uplink and downlink communication process. Unlike FDD LTE, TDD LTE can assign appropriate portions of total bandwidth to the uplink and downlink transmissions by selecting an appropriate TDD UL/DL configuration depending on the uplink/downlink traffic load. However, a fixed TDD UL/DL configuration may not be optimal for a cell due to time-varying traffic asymmetry between the uplink and downlink traffic loads. Thus, a dynamic TDD UL/DL reconfiguration mechanism is needed that adapts to the prevailing load conditions in both uplink and downlink. The signaling for dynamic TDD UL/DL reconfiguration can be supported by utilizing dual connectivity in Release 12 where a macrocell is used to coordinate the UL/DL reconfiguration for small cells.

Dynamic TDD UL/DL reconfiguration may however cause new types of inter-cell interference conditions such as eNodeB-to-eNodeB and UE-to-UE interference. This occurs in subframes that are configured differently (UL or DL) in adjacent cells. In LTE TDD some subframes are *fixed* (either UL, DL, or special) and certain subframes are *flexible*, which can be both UL or DL depending on the TDD UL/DL configuration. This is shown in Figure 11.1.

Heterogeneous Networks in LTE-Advanced, First Edition. Joydeep Acharya, Long Gao and Sudhanshu Gaur.
© 2014 John Wiley & Sons, Ltd. Published 2014 by John Wiley & Sons, Ltd.
Companion Wesite: www.ltehetnet.com.

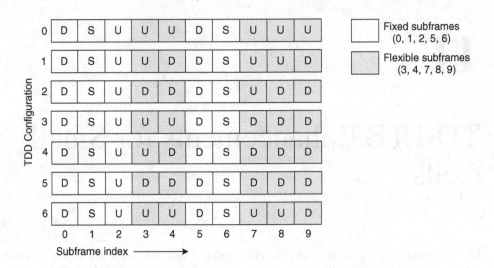

Figure 11.1 Fixed and flexible subframes in various TDD UL/DL configurations

These new interference situations therefore occur in conflicting flexible subframes and necessitate the development of new interference mitigation techniques. This chapter discusses the target deployment scenarios and potential interference mitigation schemes for dynamic TDD UL/DL reconfigurations [1].

11.1.1 TDD UL/DL Reconfiguration Scenarios in 3GPP

Extensive simulations were conducted in 3GPP to identify the deployment scenarios that are likely to benefit from TDD UL/DL reconfiguration. It was agreed not to consider macrocells for dynamic UL/DL reconfiguration as the high transmit power used in the downlink of macrocells and the high LOS probability of the channel between neighboring macro eNodeBs leads to severe eNodeB-to-eNodeB interference. This will offset any performance gains of dynamic TDD UL/DL reconfiguration. On the other hand, use of dynamic TDD UL/DL reconfiguration in small cells is promising because of the low transmission power and high channel attenuation between small cells. In addition, small cells cater to localized traffic and are more likely to witness rapid fluctuations in both uplink and downlink traffic. Dynamic TDD UL/DL reconfiguration is therefore more relevant for small cells. Based on the extensive performance evaluations in RAN1 and RAN4, the four deployment scenarios described in the following sections have been identified for applying different UL/DL configurations in small cells [1].

11.1.1.1 Scenario 1: Isolated Small Cell

This scenario provides an upper bound for throughput gains achievable via TDD UL/DL reconfiguration due to the absence of inter-cell interference. An isolated small cell can dynamically adjust the TDD UL/DL configuration to match the traffic fluctuations in the uplink and downlink to improve throughput performance. Figure 11.2 shows a scenario where the average downlink traffic arrival ratio is twice that of the uplink. The instantaneous uplink and downlink traffic load in the system depends on the average traffic arrival rates, the channel conditions, and the scheduler. Figure 11.2 compares the average throughput performance of fixed TDD

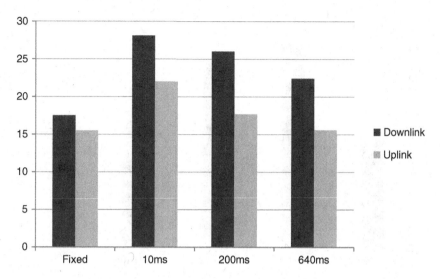

Figure 11.2 Cell average packet throughput for Scenario 1 under different timescales for TDD reconfiguration with fixed DL/UL traffic arrival ratio of 2:1. (Source: NTT DoCoMo 2013 [2]. Reproduced with permission of ETSI.)

configuration with that of dynamic TDD configurations with different reconfiguration signaling mechanisms. They are system information signaling (~640 ms), higher layer signaling (~200 ms), and physical layer signaling (~10 ms). As seen from Figure 11.2, faster dynamic UL/DL reconfiguration can provide over 40% and 60% gains for uplink and downlink data rates, respectively.

11.1.1.2 Scenario 2: Outdoor Small Cells Operated on Different Bands Than Macro

This scenario refers to the deployment of multiple small cells that either operate on a different frequency band from macrocells or are without macro coverage. For example, macrocells can operate legacy carrier (2 GHz) whereas small cells are operated on higher frequency bands such as 3.5 GHz. In this scenario, inter-cell UL/DL interference may exist if small cells do not coordinate while selecting their individual TDD UL/DL configurations. However, the potentially more severe interference between macro and small cells is not present. Figure 11.3 shows the average throughput performances for dynamic TDD without any interference mitigation mechanisms. It can be observed that while TDD reconfiguration boosts downlink data rates by over 75%, the uplink data rates do not see any gain even with the fastest timescales for TDD reconfiguration. This highlights the need for inter-cell interference mitigation mechanisms.

11.1.1.3 Scenario 3: Adjacent-channel Outdoor Small Cells Within Macro Coverage

In this scenario, small cells and macrocells operate on the same frequency band but in different carriers. Unlike Scenario 2, some adjacent channel interference between macro and small cells may exist due to the use of conflicting TDD configurations. Extensive evaluations show that dynamic TDD can offer gains in the small cell downlink and uplink data rates for low traffic load case [1].

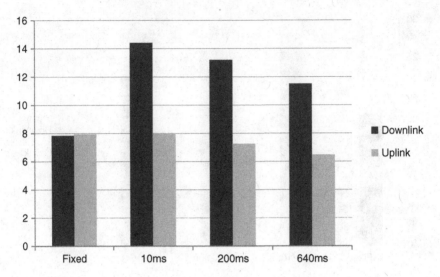

Figure 11.3   Cell average packet throughput for Scenario 2 under different timescales for TDD reconfiguration with fixed UL/DL traffic arrival ratio of 2:1. (Source: NTT DoCoMo 2013 [2]. Reproduced with permission of ETSI.)

11.1.1.4   Scenario 4: Co-channel Outdoor Small Cells Within Macro Coverage

This scenario is the most challenging as the macrocell and the small cells are all deployed on the same carrier. For instance, severe interference is caused in the uplink reception of small cells in the conflicting flexible subframes which are configured as *downlink* for macrocell and *uplink* for small cells. This eNodeB-to-eNodeB interference is very severe due to the high transmission power of the macrocell as well as LOS propagation conditions between macro eNodeB and small cell nodes. On the other hand, since small cells have a high probability of selecting downlink-centric TDD configuration (i.e. they choose TDD configuration whose flexible subframes are mostly configured as downlink), downlink transmission of small cells will also cause severe interference to the uplink transmission of the macrocell. In addition, several challenging interference situations may arise when the uplink transmission of the macro UE interferes with a small cell's uplink or downlink transmission. This happens especially when the macro UE is located near the cell edge and uses high transmit power to compensate for a large path loss to the macro eNodeB. Evaluations [1] show significantly decreased uplink throughput for both macro and small cells as any gains due to dynamic UL/DL reconfiguration are offset by severe interference due to the opposite direction of transmission in some other cells. Efficient interference mitigation schemes are therefore necessary to realize throughput gains with dynamic TDD UL/DL reconfiguration in small cells.

11.1.2   Interference Mitigation Schemes

As discussed in the previous section, dynamic TDD UL/DL reconfiguration may lead to severe interference in conflicting flexible subframes of the neighboring cells that use different TDD

UL/DL configurations. New types of inter-cell interference situations that arise due to dynamic TDD UL/DL reconfiguration are listed in the following.

- *eNodeB–eNodeB interference*: This arises when an uplink reception by an eNodeB is interfered by the downlink transmission of a neighboring eNodeB. It is the most severe inter-cell interference as eNodeBs not only transmit with maximum power but also usually have a strong coupling due to LOS propagation conditions. These factors, along with relatively low-powered uplink transmission by a UE, leads to poor uplink SINR.
- *UE–UE interference*: This arises when the downlink reception by a UE is interfered by an uplink transmission of a neighboring cell's UE. The impact on downlink SINR is more pronounced for the cell-edge UEs as the received downlink signal strength is relatively poor compared to the interference from high-powered uplink transmission by a neighboring cell's cell-edge UE.

Figures 11.4 and 11.5 illustrate the new interference situations induced by dynamic TDD UL/DL reconfiguration in Scenario 2 and Scenario 4, respectively. Among these interference situations, the UE–UE interference is not as critical as eNodeB–eNodeB interference due to the low probability of two nearby UEs belonging to different cells and having transmissions in the opposite directions. As a result, eNodeB–eNodeB interference is likely to be a dominating factor that will limit the overall benefits of dynamic TDD UL/DL reconfiguration for

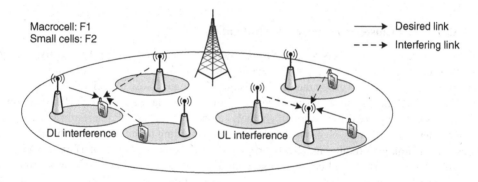

Figure 11.4 Downlink and uplink interference situations in flexible subframes for Scenario 2

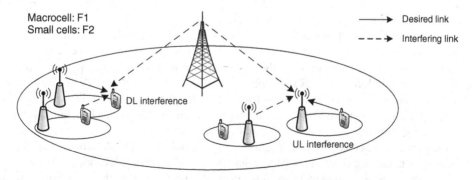

Figure 11.5 An example of interference situations in flexible subframes for Scenario 4

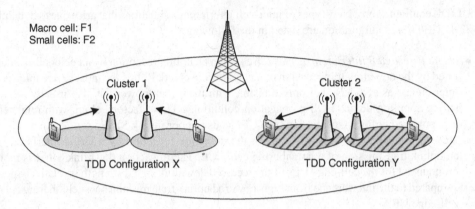

Figure 11.6 Illustration of cell clustering approach for interference management for dynamic TDD reconfiguration

traffic adaptation. New interference mitigation schemes are therefore needed to improve the packet throughput performance with dynamic TDD UL/DL reconfiguration. For the previously mentioned deployment scenarios [1], Release 12 discussions in 3GPP have focused on the interference mitigation mechanisms described in the following sections.

11.1.2.1 Cell Clustering Interference Mitigation (CCIM)

The simplest way to avoid eNodeB–eNodeB interference in small cells is to identify isolated cells and restrict dynamic TDD UL/DL reconfiguration to these cells only. A further enhancement of this scheme is to group the small cells that are located close to each other into clusters and ensure that the transmissions in all cells within a cluster have the same UL/DL configuration. This will help in the mitigation of eNodeB–eNodeB and UE–UE interference within the cell cluster. The cells may be grouped into a cluster according to the long-term channel statistics such as link gain (coupling loss) between the small cells or mutual interference between neighboring cells. If a small cell has large coupling losses to all its neighboring small cells, then it can be seen as an isolated cell and allowed to autonomously select the TDD UL/DL configuration to match its own traffic fluctuations in uplink and downlink directions. On the other hand, if it has relatively low coupling losses with some of the neighboring small cells then it can be clustered together with those neighboring cells.

As seen from Figure 11.6, a cell cluster can comprise one or more cells and isolated cell clusters may be allowed to use different TDD UL/DL configurations. For each cell cluster, a centralized node can be used to select the TDD configuration based on the periodic reports of uplink and downlink traffic loads from the cells in the cluster. If the small cells are in macro coverage, then macro is an ideal choice for the coordination of TDD UL/DL configuration, especially using the Release 12 dual connectivity mechanism. CCIM can be applied for interference mitigation in Scenario 2 as shown in Figure 11.6. However, the interference generated from macrocells in Scenarios 3 and 4 cannot be countered by CCIM. Additional interference mitigation schemes are therefore needed to handle such interference scenarios. Also note that while CCIM is easier to implement and transparent to a UE, it restricts the gains from dynamic TDD UL/DL reconfiguration as it imposes the same TDD configuration in all cells within the cluster irrespective of their traffic load conditions.

11.1.2.2 Scheduling-Dependent Interference Mitigation (SDIM)

An eNodeB can also adjust its scheduling strategies to counter the inter-cell interference induced by dynamic TDD UL/DL reconfiguration. The scheduling strategies can include link adaptation, resource allocation, transmit power control, and selection of TDD UL/DL configuration considering its own traffic requirements as well as the prevailing inter-cell interference conditions. The adjustment of scheduling strategies can be based on the observed inter-cell interference as well as the estimation of induced interference. As seen in Figure 11.1, all or a subset of flexible subframes are likely to be affected by eNodeB–eNodeB and/or UE–UE interferences. The scheduling strategies can therefore be different for subframes with different interference situations. For example, high-MCS can be used for subframes that are marginally affected by interference whereas a lower MCS needs to be used for subframes with significant interference levels. In an extreme interference situation, no resource allocation should be done. Furthermore, an eNodeB which has a tight coupling with neighboring eNodeBs can avoid scheduling cell-edge UEs for uplink transmission in the interfered subframes or perform uplink power control to mitigate eNodeB–eNodeB inter-cell interference. Likewise, in order to reduce the induced interference to neighboring cells, an eNodeB can instruct its UEs to reduce the uplink transmission power in PRBs that suffer from high interference, or completely avoid scheduling UEs in these PRBs for uplink transmission.

The interference information needed by SDIM can be obtained by eNodeB measurements, UE measurements or inter-eNodeB information exchange over X2 interface or over the air interface. The interference measurements in the downlink subframes can be divided into multiple measurement sets to account for different interference situations on flexible and fixed downlink subframes based on TDD configurations used in neighboring cells. For example, a UE can perform separate measurements for subframes with or without UE–UE inter-cell interference via the restricted measurement approach that was introduced in eICIC (see Section 5.3.2). The uplink interference needs to be measured at the eNodeB. Similar to the measurement in the downlink, the uplink measurements can be divided into multiple measurement sets according to the prevailing inter-eNodeB interference situation.

SDIM with interference measurements over the air interface is more dynamic than CCIM which requires a centralized node to coordinate TDD UL/DL configuration. If the small cells are not tightly coupled, then SDIM will achieve better performance than CCIM since CCIM reduces the traffic adaptation capabilities within the cluster, especially when the number of cells in a cluster is larger. However, with strong eNodeB–eNodeB and/or UE–UE interferences, CCIM can be more effective than SDIM. It may be more beneficial to jointly utilize CCIM and SDIM given their advantages for different interference scenarios.

11.1.2.3 Enhancement of Legacy ICIC Mechanisms

Several useful inter-cell interference mitigation methods have been specified in LTE Releases 8, 10, and 11 such as frequency-domain interference information exchange via X2 and time-domain-based ICIC with ABS. However, these methods were developed for conventional inter-cell interference situations such as eNodeB–UE interferences. If they are to be used for the new interference situations that arise from dynamic TDD UL/DL reconfigurations, they require the support of additional signaling. For example, existing Release 8 ICIC mechanism enables a victim eNodeB to send OI to neighbor cells prompting them to avoid scheduling cell-edge UEs in PRBs indicated to be strongly interfered. Now with additional eNodeB–eNodeB interference, the neighbor cell that receives OI will not know whether the high interference is due to its uplink transmission or the downlink transmission. New signaling enhancements are therefore needed to ensure that legacy ICIC mechanisms effectively deal with inter-cell interferences induced by dynamic TDD.

11.1.2.4 Enhancement of Legacy Power Control Mechanisms

Transmit power control (TPC) is another legacy mechanism that can be useful for mitigating eNodeB–eNodeB interference induced by dynamic TDD reconfigurations in small cells. For instance, a victim small cell can utilize uplink power control mechanism to instruct a scheduled UE to increase its transmit power on conflicting flexible subframes to overcome the uplink interference from neighboring cells. However, legacy TPC mechanisms may not be sufficient as the interference conditions on fixed and flexible subframes are very different. Furthermore, the interference levels on flexible subframes may also be very different depending upon the TDD configurations used by neighboring cells.

As an example, Figure 11.7 shows a victim cell with 3 different interference levels for its uplink subframes. Enhancement of legacy TPC mechanisms is required to counter these interference situations. As seen from Figure 11.8, enhanced uplink transmit power control mechanisms can yield significant throughput gains over fixed TDD configuration and even outperform CCIM. However, note that UE transmit power increase may not always be enough to counter eNodeB–eNodeB interference especially when small cells are tightly coupled. Additionally, uplink TPC may inject additional interference in the network causing increased UE–UE interference as well as higher UE energy consumption.

Apart from uplink power control, the downlink power control can also be useful in mitigating eNodeB–eNodeB interference where the aggressor eNodeB can either reduce its

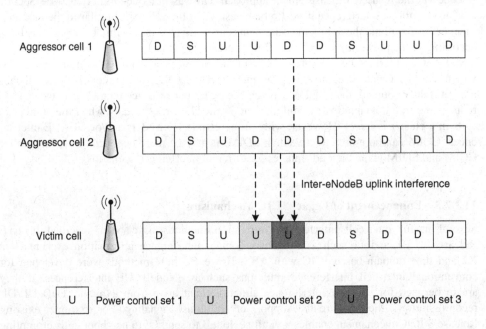

Figure 11.7 Illustration of uplink power control for countering inter-cell interference

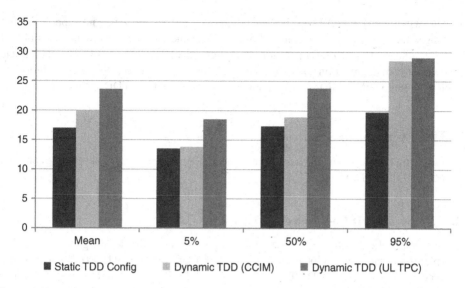

Figure 11.8 Comparison of the performances of CCIM and uplink TPC for Scenario 4 with TDD reconfiguration period of 10 ms. (Source: ZTE 2013 [3]. Reproduced with permission of ETSI.)

transmit power or avoid transmissions altogether in certain conflicting flexible subframes. However, significant power reduction may be needed for tightly coupled small cells which can cause increased UE–UE interference. The reduction of downlink transmit power will also have the undesirable effect of reduced cell coverage for subframes where downlink power control is performed.

11.1.2.5 Interference Suppressing Interference Mitigation (ISIM)

ISIM is an eNodeB receiver implementation solution for reducing inter-cell eNodeB–eNodeB interference. An eNodeB may employ an advanced receiver capable of suppressing one or more dominant interfering signals arising from its neighboring cells. Use of MMSE interference rejection and cancellation (IRC) receiver is one example of ISIM solution. Additional signaling may be needed for assisting the interference suppression. This includes the network coordination information and measurements from the eNodeB/UE. Note that ISIM can work with or without other interference mitigation schemes.

11.2 FDD-TDD Joint Operation

LTE Release 12 also initiated the interesting possibility of the joint operation of FDD and TDD carriers. Joint FDD/TDD operation is relevant for situations where a cellular operator owns both FDD and TDD spectra in the same geographical area. It will therefore be desirable to have mechanisms that allow UEs to connect to multiple carriers with different duplex modes (TDD or FDD) for improving spectral efficiency, load balancing among the carriers, and the user throughput. One possible rudimentary internetworking solution between FDD and TDD is already specified in LTE Releases 8–11. This is called the *dual mode* UE operation where a

UE can switch between the TDD and FDD duplexing modes. However, no support exists for allowing simultaneous UE operation on both the FDD and TDD carriers. Other possible solutions to enable joint TDD-FDD operation are carrier aggregation between TDD and FDD carriers, multi-stream aggregation and enhanced dual mode. Before describing these solutions, the key deployment scenarios that support joint operation on FDD and TDD carriers are explained.

11.2.1 Deployment Scenarios

The following two deployments have been considered in Release 12 for supporting joint operation on FDD and TDD bands [4].

11.2.1.1 Scenario 1: Co-Located FDD and TDD Carriers

In this scenario, FDD and TDD carriers are used at the same eNodeB. Figure 11.9 shows an example of overlaid and co-located deployment of FDD (F1) and TDD (F2) carriers. Here, the eNodeB serves legacy FDD UEs on F1, legacy TDD UEs on F2, and the joint FDD-TDD-capable UEs on both the carriers. In order to allow joint FDD-TDD operation, existing CA mechanisms of LTE Release 10/11 (CA deployment scenarios 1, 2, and 3) may be utilized with some enhancements to account for different duplexing modes of the aggregating carriers.

11.2.1.2 Scenario 2: Non-Co-Located FDD and TDD Carriers

In this scenario, the FDD and TDD carriers are located at different eNodeBs. Figure 11.10 shows an example macro-pico deployment where the FDD (F1) and TDD (F2) carriers are deployed at the macro- and picocells, respectively. In this example, legacy FDD UEs are served by the macrocell, legacy TDD UEs are served by the picocell, and the joint FDD-TDD-capable UEs are served by both the eNodeBs simultaneously. The LTE backhaul quality connecting the macro and pico eNodeBs can also lead to markedly different requirements and potential solutions for supporting joint FDD-TDD operation. If ideal backhaul is considered, then TDD-FDD carrier aggregation-based solution may possibly be based on CA

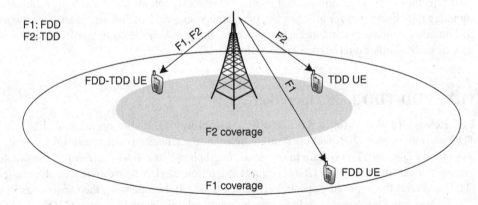

Figure 11.9   An example deployment with co-located FDD and TDD carriers

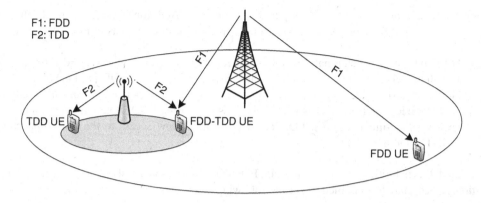

Figure 11.10   An example heterogeneous deployment with non-co-located FDD and TDD carriers

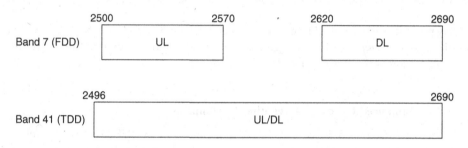

Figure 11.11   An example of overlapping FDD and TDD bands

deployment Scenario 4. However, if a non-ideal backhaul is considered, then other solutions such as multi-stream aggregation (MSA) and enhanced dual-mode (EDM) operation may be considered to allow joint operation on FDD and TDD spectra. These are explained in the following section.

11.2.2   Issues and Potential Solutions

In the existing deployments of FDD and TDD carriers, the TDD spectrum lies in similar frequency bands compared to the FDD spectrum. New spectra in high-frequency bands will become available in the future, however. These will mostly be unpaired spectra where only TDD can be deployed. The TDD carrier frequency may be separated (from adjacent channel interference point of view) from the FDD UL and DL bands, be close to either band, or even be located in between the two. It is also possible that the FDD and TDD bands overlap as shown in Figure 11.11. For this situation it is also possible that TDD can be operated in the uplink channel of a paired frequency allocation. What this means is that if an operator who has deployed FDD observes that there is hardly any FDD traffic in the uplink, he can utilize that spectrum for TDD operations. The design of joint operation on FDD-TDD bands needs to take all these combinations and flexibilities into account. Additionally, any potential solution for joint FDD-TDD network operation should satisfy a list of criteria as follows [4].

- The FDD carrier which is part of a jointly operated FDD-TDD network should allow access to both the legacy FDD UEs as well as the UEs supporting FDD-TDD joint operation.
- The TDD carrier which is part of a jointly operated FDD-TDD network should allow access to both the legacy TDD UEs as well as the UEs supporting FDD-TDD joint operation.
- Legacy single-mode carriers, FDD or TDD, should allow access to the UEs supporting FDD-TDD joint operation.

The UE capabilities for supporting joint FDD-TDD operation are also important considerations. These may be classified as per the following categories.

- Simultaneous reception on FDD and TDD carriers: this is also known as DL aggregation.
- Simultaneous transmission on FDD and TDD carriers: this is also known as UL aggregation.
- Simultaneous reception and transmission on FDD and TDD carriers.

Any potential solution should support UEs with one or more of these capabilities [4]. Based on these aspects, Release 12 considered the potential solutions described in the following sections for joint operation on FDD and TDD bands.

11.2.2.1 Enhanced Carrier Aggregation Mechanisms

As discussed in Chapter 5, Release 10/11 CA mechanisms allow a UE to access multiple carriers simultaneously if the same duplex mode is used in all the carriers. The multiple carriers can be co-located at the same eNodeB or be located at different eNodeBs that are connected via ideal backhaul. As a straightforward extension of Release 10/11 CA, the FDD-TDD CA may be used to allow data transmission on both the FDD and the TDD carriers [5, 6, 7]. The main specification impact lies in the physical layer, particularly the downlink and uplink HARQ process management. For example, for legacy TDD carrier, the HARQ information corresponding to multiple DL subframes are aggregated and transmitted in one UL subframe. The exact nature of the mapping between multiple DL subframes and one UL subframe depends on the TDD configuration. For TDD-FDD joint operation with a FDD PCell and TDD SCell, the DL subframes are in the TDD carrier but there is always a corresponding UL subframe in the FDD carrier. The HARQ mechanism for TDD may therefore be changed.

To summarize, the design of HARQ timing depends on two factors as follows.

- *Uplink CA capability*: For UEs supporting uplink CA, it is possible to have independent HARQ feedbacks for each serving cell. For this situation the existing HARQ timing mechanism can be reused. For UEs not supporting uplink CA, HARQ feedback needs to be transmitted on a single carrier.
- *The duplex mode of PCell*: In the case where a FDD carrier is configured as a PCell, the existing HARQ timing for FDD can be reused. A new HARQ timeline is likely to be required when a TDD carrier is configured as PCell. For a FDD PCell and a TDD SCell, cross-carrier scheduling of PDSCH cannot apply in UL subframes in the TDD SCell (likewise for PUSCH transmission in DL subframes), leading to reduction in gains of CA. For a TDD PCell and

a FDD SCell, since the PCell has a smaller number of downlink subframes than the SCell, multi/cross-subframe scheduling is necessary for scheduling the additional subframes in the FDD SCell; otherwise, a significant percentage of subframes will be wasted.

11.2.2.2 Enhanced Dual-Mode Operation

Dual-mode operation has been specified in LTE Releases 8–11 in which a UE is capable of operating on a FDD serving cell and also a TDD serving cell. The network performs handover to switch a UE between the FDD serving cell and the TDD serving cell. This is typically a slow process. Enhanced dual-mode [7] operation was introduced in the LTE Release 12 standardization phase to support simultaneous operations on the duplexing modes. An enhanced dual-mode UE refers to the situation where a dual-mode UE can connect to both the FDD serving cell and the TDD serving cell, and the switch between the serving cells with different duplex modes is carried out at a faster timescale than conventional handover. Similar to a conventional dual-mode operation, it can only receive or transmit on one serving cell in a subframe.

Enhanced dual-mode UEs can independently access a FDD serving cell and a TDD serving cell without the network knowing the capabilities of the UE. However, this will require more RF and baseband capabilities at the UE. Processing at the UE will be reduced if the network has knowledge of the UE capability. High-layer specification changes may be required to support enhanced dual mode. For example, the UE may need to establish two RRC connections with the network, one for each serving cell.

References

[1] 3GPP (2012) Further enhancements to LTE Time Division Duplex (TDD) for Downlink-Uplink (DL-UL) interference management and traffic adaptation 3GPP TR 36.828 (Release 11). Third Generation Partnership Project, Technical Report, June 2012.
[2] NTT DoCoMo (2013) R1-130755, deployment scenarios and interference mitigation schemes for eIMTA. NTT DoCoMo, Technical Report, January 2013.
[3] ZTE (2013) R1-130128, power control enhancement for performance improvement in multi-cell scenarios. ZTE Technical Report, January 2013.
[4] 3GPP (2013) Evolved Universal Terrestrial Radio Access (E-UTRA), Study on LTE TDD-FDD Joint Operation including Carrier Aggregation TR 36.847 v0.1.0. Third Generation Partnership Project, Technical Report, August 2013.
[5] CATT (2013) R1-133022, Potential Solutions for LTE FDD-TDD Joint Operation. CATT, Technical Report, August 2013.
[6] Samsung (2013) R1-133102, Solutions for FDD-TDD Joint Operation. Samsung, Technical Report, August 2013.
[7] Huawei (2013) R1-132886, Potential Solutions of TDD-FDD Joint Operation. Huawei, Technical Report, August 2013.

12

Full Dimension MIMO

12.1 Introduction

Dense deployment of small cells increases spatial reuse and can improve network through-put, if interference is managed properly. Given the complex nature of network deployments and ever-rising data demand however, additional technologies are needed to complement the gains from installing extra small cells. One particularly promising technology which further enhances spatial reuse is called Full Dimension MIMO (FD-MIMO) [1]. In this technology, Active Antenna Systems (AAS) [2] are used that enable dynamic three-dimensional shap-ing of the beams that are transmitted by the base stations. As will be shown subsequently, FD-MIMO using AAS increases capacity and coverage, enables more flexible deployment, and reduces operator cost. FD-MIMO techniques are especially helpful for urban, high-rise scenarios which will generate the maximum amount of data demand in the future.

The benefits of FD-MIMO are best understood by contrasting the characteristics of AAS with those of passive antenna systems. The following sections explain these in detail.

12.2 Antenna Systems Architecture: Passive and Active

The high-level functional blocks of a base station are shown in Figure 12.1. They consist of the baseband unit, the transceiver and duplexer unit, and the antenna components. The baseband unit is responsible for the digital processing operations. The digital front end shapes the sig-nal to avoid subsequent non-linearities in the power amplifier which can happen for multiple reasons such as high PAPR. It also alters the sampling rate to make the signal compatible with the rate of the Digital-to-Analog/Analog-to-Digital Converters. The RF unit consists of com-ponents such as power amplifiers and duplexers which are analog filters tuned to the center frequencies of the signals of interest that enable simultaneous transmission and reception over the same set of physical antennae for FDD. In this section the combination of the RF units and the Digital Front End units are referred to as the Transceiver Unit or the Transmit Receive Duplexer Unit (TRDU). A TRDU is thus an active unit with the capability to transmit and/or receive radio signals.

Base station architectures have evolved recently [3], as shown in Figure 12.2. For traditional base stations with passive antenna systems, the baseband and the TRDU are located at the

Heterogeneous Networks in LTE-Advanced, First Edition. Joydeep Acharya, Long Gao and Sudhanshu Gaur.
© 2014 John Wiley & Sons, Ltd. Published 2014 by John Wiley & Sons, Ltd.
Companion Wesite: www.ltehetnet.com.

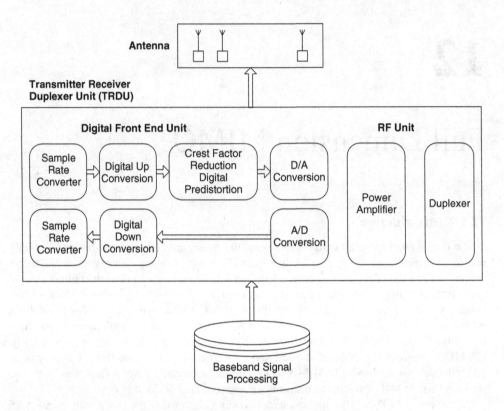

Figure 12.1   Functional blocks of a base station

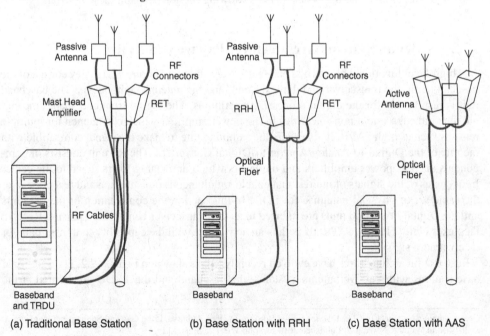

Figure 12.2   Evolution of base station architectures from passive to active antenna based

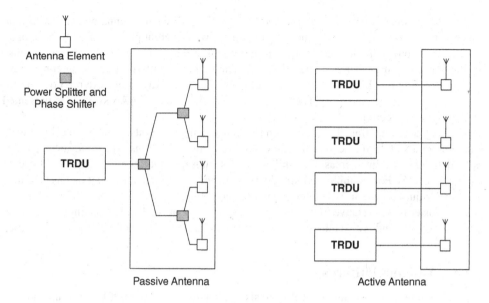

Figure 12.3 Features of passive and active antenna systems

base of the cell tower and numerous RF cables are fixed along the body of the cell tower for transporting the RF signals to the antennae at the top (also called masthead). The antennae are said to be *passive* as they simply radiate the radio signals that are transmitted to them. The RF cables are very bulky and waste a lot of power in transmitting the signal up to the masthead. Site engineering costs that the operators have to bear, such as rental fees and labor costs for installation, are high.

Some of the drawbacks of a traditional base station installation can be remedied by using remote radio heads (RRH) as shown in Figure 12.2. In this design the RRH units, which essentially perform the TRDU functionalities, are separated from the baseband unit and located close to the antennae. The baseband output is connected to the RRHs via a single optical fiber which transports signals using the Common Public Radio Interface (CPRI) or the Open Base Station Architecture Initiative (OBSAI) protocols. There is much less loss in the optical fiber as compared to the RF cables. The site engineering costs are also much lower [4, 5]. There is still a small RF connector cable that connects the outputs of the RRHs to the antennae.

Both traditional base stations and those with RRH use passive antenna systems. The AAS architecture goes one step beyond RRH-based antenna systems and integrates the antenna into the TRDU. This combination is called an active antenna. The difference between passive and active antennas is depicted in Figure 12.3.

As shown in Figure 12.3 an antenna (or, more correctly, an antenna port) consists of a number of individual antenna elements. This is a usual practice in cellular systems. The individual antenna elements are also called the radiating elements of the antenna port. As shown in the figure, there is one TRDU per passive antenna port. A passive radio distribution network divides the output of the TRDU and feeds power to each antenna element. This usually requires the output of the power amplifier in the TRDU to be high. The radio distribution network also applies remote electrical tilt (RET) to the antenna waveform to steer the direction of its maximum gain. This is performed by an electromechanical actuator that applies phase shifts to the input signals of each antenna element.

By contrast, in AAS there is a corresponding TRDU for each antenna element within an antenna port. The antenna port is now said to be *active*. Different phase shifts can be applied to the inputs on each antenna element in the digital domain to steer the resultant response of the port along the desired spatial direction. The radiation pattern of active antenna arrays can therefore be flexibly adjusted and dynamically altered in ways that were not possible for RET-enabled passive antennae. The mathematical formulation of AAS systems is presented in the following section.

In addition to enabling flexible antenna responses, the AAS also enables the increasing miniaturization of the transceiver and radio units and more operations to be performed in a software defined way. This increases flexibility and ease of implementation while further reducing operator cost [5]. For example, an individual TRDU for each antenna element means that there is an individual power amplifier for each antenna element. Each power amplifier will transmit at a lower power relative to the single power amplifier of the RRH. This improves overall power efficiency and is also inherently more robust to failures of an individual power amplifier.

12.3 Antenna Patterns

This section presents a mathematical analysis of the radiation pattern of a single antenna element and the combined pattern of an array.

12.3.1 Passive Antenna Element Pattern

Let us start with the conventional passive antenna system. The reference three-dimensional coordinate system is shown in Figure 12.4. The azimuth angle ϕ varies over the range $[-180°, 180°]$ with $\phi = 0$ along the x-axis. The elevation angle θ varies within the range $[0°, 180°]$ with $\theta = 0$ corresponding to the vertical direction (zenith) and $\theta = 90°$ being parallel to the x–y plane. The elevation angle θ is also sometimes referred to as the *zenith angle*. We denote the radiation pattern of a single antenna element $U(\theta, \phi)$.[1] For a wide range of antennae used in practical cellular systems, the antenna element pattern can be approximated by the sum of two one-dimensional patterns [7], namely $A_H(\phi)$ which is in the azimuth (horizontal) direction and $A_V(\theta)$ in the elevation (vertical) direction. These patterns are defined [2]:

$$10 \log (U(\theta, \phi)) = G_{max, dB} - \min\{-[A_H(\phi) + A_V(\theta)], A_m\} \tag{12.1}$$

$$A_H(\phi) = -\min\left[12\left(\frac{\phi - \phi_{sect}}{\phi_{3dB}}\right)^2, SLL_h\right] \tag{12.2}$$

$$A_V(\theta) = -\min\left[12\left(\frac{\theta - \theta_{RET}}{\theta_{3dB}}\right)^2, SLL_v\right]. \tag{12.3}$$

In the above equations SLL_h and SLL_v represent the side-lobe levels of the respective radiation patterns. The values of ϕ_{3dB} and θ_{3dB} represent the half-power bandwidths of

---

[1] If $P_E(\theta, \phi)$ is the magnitude of the antenna element pattern, the radiation pattern (also called radiation intensity) is given by $U(\theta, \phi) = |P_E(\theta, \phi)|^2$ [6, chapter 16]. If the maximum value of the radiation pattern is G_{max}, then the power gain of the antenna element is given by $G(\theta, \phi) = U(\theta, \phi)/G_{max}$.

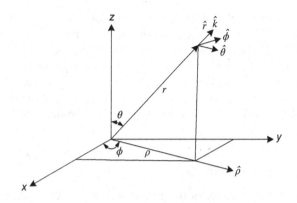

Figure 12.4 Cartesian, cylindrical and spherical coordinates

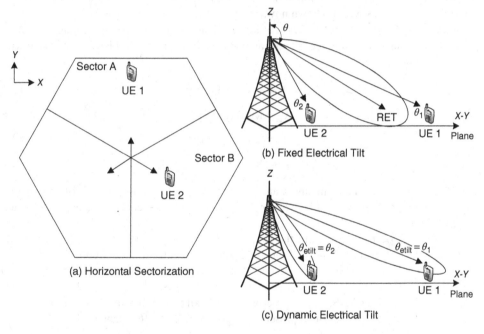

Figure 12.5 Example layout showing (a) horizontal sectors and elevation beamforming with (b) fixed RET and (c) dynamic electrical tilts

the horizontal and vertical radiation patterns. The value ϕ_{sect} is used to denote the sector and is equal to the angle pointing to the broadside (main lobe) of the sector. For example, $\phi_{sect} = 90°, 90° + 120°, 90° + 2 \times 120°$ for three sectors as per the layout depicted in Figure 12.5(a). The notation θ_{RET} represents the remote electrical tilt of the base station antenna ports.

First note that the use of three different values of ϕ_{sect} leads to the creation of the three sectors in the azimuth domain. This is referred to as *beamforming* along the azimuth direction, with three fixed beams that are 120° apart. In the elevation domain, only one beam is possible: the beam whose broadside is along the tilt direction θ_{RET}. All these different beams are oriented along fixed directions in the azimuth or elevation directions and cannot be changed dynamically. This situation is illustrated in Figure 12.5(a) and (b). In the figure, the two UEs can be served by separate beams in the horizontal direction, i.e. UE 1 is associated with the base station in sector A and UE 2 is associated with the base station in sector B. In the elevation dimension the broadside points in the direction θ_{RET}, while the elevation angles of UEs 1 and 2 are θ_1 and θ_2.

12.3.2   Active Antenna Systems

Figure 12.5(a) shows the existence of three fixed horizontal sectors while 12.5(b) shows the single vertical sector characterized by the value of the remote electrical tilt. A natural question that arises is whether it is possible to form multiple vertical sectors having different values of electrical tilts. A possible depiction is shown in Figure 12.5(c). Such dynamic-tilting is indeed possible with AAS as will be shown now.

Figure 12.4 shows the wave vector $\vec{k} = \mathbf{k}\hat{k}$ that denotes the direction of propagation of the radio waves transmitted from the antenna array where $\mathbf{k} = 2\pi/\lambda$. For this vector the following properties hold:

$$\vec{k} = k_x\hat{x} + k_y\hat{y} + k_z\hat{z} \tag{12.4}$$

$$k_x = \mathbf{k}\cos(\phi)\sin(\theta) \tag{12.5}$$

$$k_y = \mathbf{k}\sin(\phi)\sin(\theta) \tag{12.6}$$

$$k_z = \mathbf{k}\cos(\theta). \tag{12.7}$$

Let there be M antenna elements in an AAS antenna port. Let the location of antenna element m in a representative port be $d_m = x_m\hat{x} + y_m\hat{y} + z_m\hat{z}$. A typical arrangement is shown in Figure 12.6.

From the perspective of the UE, the antenna elements within a port are not visible as the same signal s is transmitted from all the antenna elements constituting a port. Specific phase shifts to s can be applied for each antenna element to steer the resultant response in a given spatial direction. Let w_m be the phase shift for element m. Recall that these phase shifts can be applied digitally using AAS, and is therefore more flexible and dynamic than RF phase shifts for RET. The *array factor* of the representative port is given by [6, chapter 19]:

$$A(\theta, \phi) = \sum_{m=0}^{M-1} w_m e^{-j\vec{k}.d_m}. \tag{12.8}$$

The overall radiation pattern of the antenna port is then given by

$$U_P(\theta, \phi) = |A(\theta, \phi)|^2 U(\theta, \phi). \tag{12.9}$$

The expression for w_m to induce beamforming in the elevation domain is

$$w_m = \frac{1}{\sqrt{M}} \exp\left(j\frac{2\pi}{\lambda}d(m-1)\cos\left(\theta_{etilt}\right)\right), \quad m = 1, \cdots, M. \tag{12.10}$$

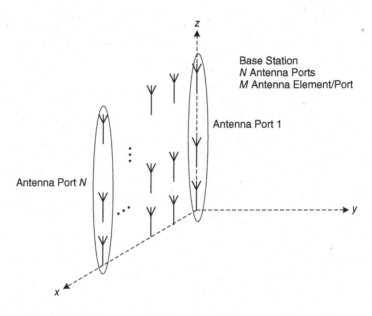

Figure 12.6   Antenna element array and element to port mapping

The electrical tilt θ_{etilt} enables dynamic elevation beamforming. This can be seen by evaluating the array factor for the representative port. For port 1, for which $d_m = (m-1)d\hat{z}$ from Figure 12.6, we can define

$$\psi = \frac{2\pi}{\lambda}d(\cos(\theta) - \cos(\theta_{etilt})), \tag{12.11}$$

from which it is easy to see that the array factor is given by

$$A(\theta, \phi) = \sum_{m=0}^{M-1} e^{-jk\psi} \tag{12.12}$$

$$= \frac{1}{\sqrt{M}}\left(\frac{1 - \exp\{-jM\psi\}}{1 - \exp\{-j\psi\}}\right) \tag{12.13}$$

$$= \frac{1}{\sqrt{M}}\frac{\exp\{-jM\psi/2\}}{\exp\{-j\psi/2\}}\left(\frac{\exp\{jM\psi/2\} - \exp\{-jM\psi/2\}}{\exp\{j\psi/2\} - \exp\{-j\psi/2\}}\right) \tag{12.14}$$

$$= \frac{1}{\sqrt{M}}\exp\left\{-j\psi\left(\frac{M-1}{2}\right)\right\}\frac{\sin(M\psi/2)}{\sin(\psi/2)}. \tag{12.15}$$

$A(\theta, \phi)$ peaks at angle $\psi = 0$ corresponding to $\theta = \theta_{etilt}$. Assume that for each antenna element pattern $U(\theta, \phi)$, the remote electrical tilt $\theta_{RET} = 90°$. By varying θ_{etilt} in the digital domain, the radiation pattern can therefore be steered towards the desired direction. This

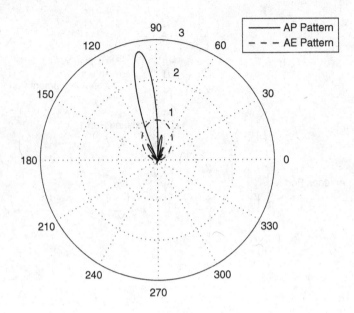

Figure 12.7 Beam steering of radiation pattern of an AAS port using electrical tilt of 100° (AE: antenna element; AP: antenna port)

phenomenon is called elevation beamforming. Figure 12.7 shows an example with four antenna elements per port, $\theta_{etilt} = 100°$ and $\theta_{RET} = 90°$. The antenna element radiation pattern has a large half-power beamwidth but that of the antenna port is focused in the direction of the electrical tilt.

Grouping of antenna elements into ports is also called *virtualization*. Other methods of virtualization than that depicted in Figure 12.6 are possible. Antenna port virtualization can be chosen to dynamically shape and steer the beam in both azimuth and elevation. To achieve this, for example, a port can now be defined as a $K \times M$ grid of antenna elements similar to the antenna grid shown in Figure 12.6. Let w_{km} be the beamforming phase weights that are applied to the antenna elements. It can be shown that by choosing

$$w_{km} = \frac{1}{\sqrt{KM}} \exp\left(j \frac{2\pi}{\lambda} d\Phi \left(\theta_{etilt}, \phi_{escan} \right) \right) \tag{12.16}$$

$$\Phi(\theta_{etilt}, \phi_{escan}) = (m-1)\cos(\theta_{etilt}) + (k-1)\sin(\theta_{etilt})\sin(\phi_{escan}) \tag{12.17}$$

$$m = 1, \cdots, M, \ k = 1, \cdots, K, \tag{12.18}$$

complete FD-MIMO beamforming along the direction $(\theta_{etilt}, \phi_{escan})$ is possible.

In general, increasing the number of elements per port increases the directivity of the response, and increasing the number of ports increases the total power and improves MIMO gains. The problem of optimally mapping a given set of antenna elements to another reduced-size set of antenna ports is open to solution.

12.3.3   AAS with Additional Mechanical Tilt

Can AAS with dynamic electrical tilt steer the resultant beam in any given direction? Unfortunately, that is not the case and the reason is not very hard to see. From Equation (12.9), the peak of the antenna element response $U(\theta, \phi)$ is at $\theta_{RET} = 90°$ and that of the antenna array response $A(\theta, \phi)$ is at $\theta = \theta_{etilt}$. Since the maximum value of $U(\theta, \phi)$ is unity and that of $A(\theta, \phi)$ is $\sqrt{M} > 1$, the combined response will still peak at $\theta = \theta_{etilt}$ (especially for large values of M) but the peak gain will reduce. It can be seen that for $|\theta_{etilt} - \theta_{RET}| \leq \theta_0$ where θ_0 is a small angular value, the gain of $U_P(\theta, \phi)$ is comparable to that of $A(\theta, \phi)$. In practice, θ_0 can be about $20°$ for $M = 8$.

If higher values of θ_{etilt} are desired, then additional mechanical tilt may be necessary. Prior to AAS and RET, traditional base stations employed mechanical tilt, i.e. physical tilting of the base station antenna array towards a given direction.

Mechanical tilt is not without its own problems. It has to be done manually and hence is cumbersome. It also distorts the antenna array responses $A_H(\theta, \phi)$ and $A_V(\theta, \phi)$. To see how, assume a global coordinate system (GCS) denoted $[\rho, \theta, \phi]$ as given in Figure 12.4. A mechanical tilt of the base station antenna array by the amount γ about the y-axis in GCS will lead to a new rotated coordinate system called the local coordinate system (LCS), denoted $[\rho, \acute{\theta}, \acute{\phi}]$. The transformation [8] is given by

$$\acute{\theta} = \cos^{-1}(\sin \gamma \sin \theta \cos \phi + \cos \gamma \cos \theta) \qquad (12.19)$$

$$\acute{\phi} = \arg(\cos \gamma \sin \theta \cos \phi - \sin \gamma \cos \theta + j \sin \theta \sin \phi). \qquad (12.20)$$

It can be seen that the expressions for $A_H(\phi)$ and $A_V(\theta)$ as given in Equation (12.1) hold in the LCS but not in GCS. The mechanical tilt can lead to pattern blooming [3], i.e. the expansion of the half-power beamwidth, leading to more inter-sector interference. A careful combination of electrical and mechanical tilts may therefore be required to achieve the desired antenna response.

12.3.4   Effect of Multipath Fading Channels

Beamforming ensures that the signal energy of the *transmit waveform* is directed towards the recipient UE whose spherical coordinates are given by the azimuth and elevation angles ϕ_L and θ_L, also called the LoS directions. But what about the *received waveform* at the UE? If there is no scattering in the channel, then the received waveform has the same directional gain as the transmit waveform. If there is random scattering (small-scale fading) however, then the signal energy is spread over multiple paths that arrive at the UE. The different multipaths have directional gains different from (ϕ_L, θ_L). Is beamforming of the transmit waveform still effective?

To understand this, consider the scattering induced by fading channels as depicted in Figure 12.8. For a wide range of realistic channels [9], it is seen that the angular spread of the multipaths are given by perturbations about ϕ_L and θ_L. Without *a priori* information about the exact nature of the multipath dispersion, it therefore makes sense for the transmitter to perform beamforming along ϕ_L and θ_L.

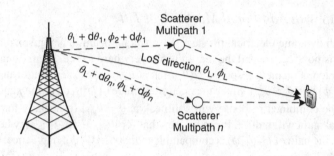

Figure 12.8   Spatial channel model for small-scale fading

Consider the frequency domain response for a $N \times 1$ MISO channel after FFT (which will also be correlated to ϕ_L and θ_L). Consider a representative subcarrier, whose response is denoted $\mathbf{h}$. The received signal at the UE is given by

$$\mathbf{y} = \sqrt{U_P(\theta_L, \phi_L)}\mathbf{h}^H \mathbf{v}s + \mathbf{z}, \tag{12.21}$$

where $\mathbf{v}$ is the precoder chosen by the UE. This could be chosen based on maximal ratio combining or singular value decomposition (SVD) based transmit precoding vector selection for rank 1 transmission to a UE with multiple antennas.

Note that $\mathbf{v}$ is also called the beamforming vector. This is not to be confused with AAS beamforming discussed previously. AAS based beamforming shapes the antenna response of each transmit port without taking explicit account of the effect of small-scale fading. SVD based beamforming aligns the transmit weights across the different antenna ports to align with the instantaneous realization of a small-scale fading channel.

12.4   FD-MIMO Deployment Scenarios

This section provides information on the various possible scenarios where FD-MIMO can be deployed [10]. Most of these scenarios are proprietary techniques used at the base station with minimal changes in the existing standard. They are broadly grouped within the categories described in the following sections.

12.4.1   UE-Specific FD-MIMO

The possibility of UE-specific beamforming was mentioned at the beginning of Section 12.3.2. This could be deployed in both macro- and small cells. In a heterogeneous network, the small cells are deployed in areas of high UE density. Due to the reduced coverage region of small cells, especially in indoor deployments, the scattering in the channel reduces. In such environments, it is easier to transmit multiple narrow beams to different UEs. Alternatively, the macro may use AAS to provide coverage in those isolated areas that otherwise do not lie within the coverage region of the small cells. Using directional beamforming avoids interference to UEs in other regions.

Also, in the case of cooperative transmission such as CoMP, it is possible that the cooperating base stations tilt their transmissions such that beams all point towards the direction of the cell-edge UE which is to be served. This increases CoMP gains.

12.4.2   Cell-Specific FD-MIMO

Tilting the transmit waveform for each individual UE may not be possible for a large number of UEs to be served and is not even necessary. For a wide range of scenarios, it may suffice to adjust the tilts to align with the location area for a group of UEs. For example, consider a small cell located in a hotspot area which only has to serve the UEs that are clustered around it. It can adjust the tilt to have a small but more focused coverage. This is depicted by *Beam 1* shown in Figure 12.9(a). If the user profile changes or the small cell wishes to serve a larger population of UEs, it can decrease its tilt to enable a larger coverage as shown by *Beam 2*. This dynamic expansion of cell ranges is achieved without the artificial CRE biasing that was implemented in Release 10 and 11 eICIC. This phenomenon is also called proactive cell shaping.

FD-MIMO can also create additional sectors along the vertical dimension, as shown in Figure 12.9(b). Both beams are simultaneously active and have different coverages in the same way as conventional horizontal sectors. The different coverage areas can therefore be viewed as different sectors (cells) and be assigned unique cell IDs. For example, the top floors of a building could be covered by *Beam 2* and assigned a different cell ID from the bottom floors which are covered by *Beam 1*.

12.4.3   System-Specific FD-MIMO

An AAS system can tune the value of electrical tilt to adjust the coverage. This fits perfectly with inter-band carrier aggregation, where operators use multiple carriers with different coverage. The higher-frequency carriers deliver a smaller coverage due to increased path loss at higher frequencies. A larger value of tilt for these carriers can be chosen in order to boost the received signal strength. Carrier-specific tilting also configures different carriers to meet different demand scenarios.

Carrier-specific tilting can also be used by operators to support different access technologies such as LTE and HSPA. Each carrier can be configured with a different access technology and

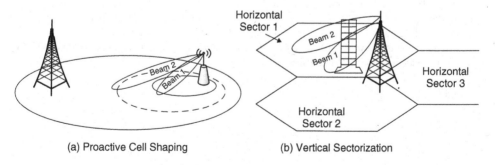

(a) Proactive Cell Shaping   (b) Vertical Sectorization

Figure 12.9   Examples depicting FD-MIMO (a) shaping the cell boundaries dynamically and (b) creating new sectors in the vertical domain

the tilt adjusted to align with the capacity and coverage demands of the specific technology. In fact, adaptive tilting is more powerful and can even be used to enable independent coexistence of different technologies in the same frequency, for example GSM and WCDMA at 900 MHz or GSM and LTE at 1800 MHz. For example, WCDMA at 900 MHz can be tuned to deliver more focused coverage than GSM at 900 MHz by tilting the WCDMA beam more than the GSM beam using the same antenna [10].

12.5 Conclusion

This chapter has discussed the characteristics of active antenna systems and their advantages over passive antenna systems. AAS enables FD-MIMO beamforming, which exploits the three-dimensional geometry of the network and UE deployments. Using FD-MIMO in conjunction with small cells has the potential to radically enable a new range of deployments and enhance their achievable capacities and coverages. Much of the work is in the initial stages and will experience increased activity in the years to come.

References

[1] Nam, Y.H., Ng, B.L., Sayana, K., Li, Y., Zhang, J., Kim, Y., and Lee, J. (2013) Full-dimension mimo (FD-MIMO) for next generation cellular technology. *IEEE Communications Magazine*, **51**(6), 172–179.

[2] 3GPP (2013) 3GPP TR 37.840, Study of AAS Base Station. Third Generation Partnership Project, Technical Report.

[3] 4G Americas (2012) MIMO and smart antennas for mobile broadband systems. 4G Americas, Technical Report, October 2012.

[4] Huawei (2012) Active antenna system: Utilizing the full potential of radio sources in the spatial domain. Huawei, White Paper.

[5] Mesaplexx (2012) The capacity crunch: is the revolution about to begin. Mesaplexx, White Paper.

[6] Orfanidis, S.J. (2008) *Electromagnetic Waves and Antennas*. Online Book, available at http://www.ece.rutgers.edu/ orfanidi/ewa/ (accessed November 2013).

[7] Seifi, N., Coldrey, M., Matthaiou, M., and Viberg, M. (2012) Impact of base station antenna tilt on the performance of network MIMO systems. In *Proceedings of IEEE Vehicular Technology Conference (VTC)*, May 2012.

[8] Fraunhofer IIS (2013) R1-134948, text proposal on mechanical and electrical antenna tilting. Fraunhofer IIS, Technical Report, October 2013.

[9] Radiocommunication Sector of ITU (2009) Report ITU-R M.2135-1, guidelines for evaluation of radio interface technologies for IMT-Advanced. ITU-R, Technical Report.

[10] Nokia Seimens Network (2012) Active antenna systems: a step-change in base station site performance. Nokia Seimens, White Paper.

13

Future Trends in Heterogeneous Networks

13.1 Summary

The impact of small-cell-based heterogeneous networks has been discussed in detail in this book. Chapter 1 described how the rapidly increasing demand for data in present times will lead to new paradigms of network operation where the focus shifts from homogeneous macrocell-based systems to a dense overlay of multiple types of small cells in a macro area. The level of interference is however magnified in a dense heterogeneous network, and efficient interference management techniques are therefore needed to harness the potential of heterogeneous networks.

LTE-based cellular systems will be the leading enablers of heterogeneous network deployments; studying the technological developments of heterogeneous networks within the framework of the 3GPP standardization is the way ahead for future innovations. Accordingly, Chapters 2–4 established the basics of the LTE standard. Wherever relevant, the underlying theory of digital wireless communications were explained and the signaling and protocol aspects of LTE Releases 8–10 were presented.

Chapters 5–7 provided a systematic study of inter-cell interference (eICIC) mechanisms that were standardized in LTE Releases 10 and 11 to mitigate the interference arising in heterogeneous networks. From simple blank subframe design and implementation, the chapters discussed more advanced transmitter and receiver signal processing mechanisms to suppress interference and improve performance. Enhancements in both data and control channel design were discussed.

Chapters 8 and 9 provide a different take on inter-cell interference by discussing the possibility of base stations being allowed to coordinate, thus transforming the energy of an interferer into useful signal energy. This technique, called CoMP, has the potential to vastly improve network performance. Several practical challenges first have to be overcome before this potential can be realized, however. The chapters presented the different CoMP categories introduced in LTE Release 11, their signal processing, and the changes introduced in Release 11 for supporting CoMP.

Heterogeneous Networks in LTE-Advanced, First Edition. Joydeep Acharya, Long Gao and Sudhanshu Gaur.
© 2014 John Wiley & Sons, Ltd. Published 2014 by John Wiley & Sons, Ltd.
Companion Wesite: www.ltehetnet.com.

Chapters 10–12 describe the state-of-the-art developments in heterogeneous networks that are currently taking place in 3GPP with the initiation of Release 12. A whole array of new technologies have been introduced such as dynamic switching of small cells, new carrier types with reduced control signaling, dynamic reconfiguration of TDD-LTE, joint configuration of TDD and FDD via carrier aggregation, and, lastly advanced MIMO signal processing with three-dimensional beamforming. All these technologies will work in unison to achieve the efficient operation of small cells.

There are many other areas of active research and development which features small cells, some of which may be adopted in future LTE standards. This final chapter provides brief summaries of some of these topics.

13.2   Small Cells and Cloud RAN

In a heterogeneous network, a large number of stand-alone eNodeBs are deployed. Each eNodeB has its own baseband processing unit, cooling equipment, back-up battery, monitoring system, and other support facilities. Such a distributed deployment has the following limitations.

1. The cost of deploying an eNodeB is high mainly due to its support facilities.
2. The interference between two neighboring eNodeBs in the same frequency is difficult to coordinate in the case of non-ideal backhaul.
3. The average utilization of the signal processing unit at each eNodeB is low due to the spatio-temporal variations in traffic. The eNodeB is designed to satisfy the peak rate requirements and is therefore under-utilized most of the times.

To address some of these issues, Cloud Radio Access Network (C-RAN) has been proposed by various carriers. The C-RAN architecture is intended to reduce operational costs by locating the baseband processing units and supporting facilities together for all eNodeBs in an area. Remote Radio Heads (RRHs) that are connected the central baseband processor via low-latency backhaul (e.g. fiber) are distributed in the area. Such network architecture also makes advanced interference management solutions such as CoMP and eICIC easier to implement due to the enhanced backhaul support. C-RAN is therefore somewhat similar to DAS architecture mentioned in Chapter 1; C-RAN is however much broader in terms of scope and area of operation (e.g. single building/stadium in case of DAS versus an entire city in the case of C-RAN).

At first glance, the centralized nature of the C-RAN philosophy seems to be in opposition to distributed processing in small cells. Recall however that no one solution will suffice in a given system. C-RAN systems require complex signal processing and storage requirements at the centralized location(s) and also fiber-based backhaul to the RRHs. C-RAN is ultimately likely to take its place beside traditional macros and emerging small cells as another tool for building cellular networks. The co-existence issues of C-RANs and small cells may prove to be an important area of future work.

13.3 Small Cells, Millimeter Wave Communications and Massive MIMO

The International Telecommunication Union (ITU) will allocate new spectra for cellular networks in 2015. It is expected that at least 1000 MHz will be assigned by 2020 in the frequency bands 1.5 GHz, 3.3–3.6 GHz, and above 5 GHz [1]. The extra spectrum will increase capacity; coverage will however be low as path loss is high at higher-frequency bands. Macrocells therefore cannot be deployed in these frequencies and small cell deployments are the only solution.

Transmission over high frequencies are referred to as *millimeter wave communications* due to the resulting small wavelengths. Millimeter waves have been used for fixed wireless communications for decades. For example, Local Multipoint Distribution Service (LMDS) operates at 28–30 GHz. Several industrial standards such as IEEE 802.11ad have discussed the adoption of multi-gigabit speed wireless communications on the unlicensed band 60 GHz. As the mobile data traffic grows, millimeter waves have also been considered as the potential candidate of next-generation (5G) cellular networks [2].

The large path loss at high frequencies reduces the spectral efficiency for millimeter wave communications, however. The key to the solution lies with a technology known as *massive MIMO*. Massive MIMO [3, 4, 5] scales up the gains of current MIMO systems by employing a very large number of antennas (100+) at the transmitter. These extra terminals enable transmission of narrow beams towards a recipient UE and reduce interference to other UEs which are spatially separated. The array gain from the large number of antennae can help offset the path loss at high frequencies. FD-MIMO with two-dimensional antenna arrays at the transmitter was discussed in Chapter 12. FD-MIMO can be viewed as the initial realization of massive MIMO within the LTE Release 12 framework with at most 80 antennae at the transmitter. In future, full-fledged massive MIMO implementations can be envisaged with a much larger number of antennae. A very large number of antennae at the base station leads to favorable propagation conditions, which means that channel response to a UE has much less probability of experiencing deep fades and channel responses of different UEs are close to orthogonal. This makes downlink MU-MIMO operations very efficient. It also means that downlink resource allocation at the MAC layer is simplified as the base stations can allocate all resources to all the UEs simultaneously and perform MU-MIMO. In the uplink, multi-user interference can be eliminated in principle with simple matched filters.

Note that the gains of massive MIMO are not dependent on the carrier frequency. For high-frequency bands, antenna elements can however be miniaturized thus avoiding bulky antenna arrays in the small cells. Massive MIMO and millimeter wave communications therefore enable each other to function.

There are other practical challenges that have to be overcome before massive MIMO becomes a reality, however. Estimating the channels for such a large number of transmit antennae is a challenge. For FDD-LTE, the overhead of CSI-RS from a large number of antenna ports can become prohibitive. TDD-LTE can solve this problem by using channel reciprocity to estimate the downlink channel from uplink reference signals transmitted by the UE. Due to the finite coherence time of the channel however, different UEs in adjacent cells may transmit on the same reference signal leading to the so-called *pilot contamination* problem. While not new to massive MIMO, the consequences of pilot contamination are more severe for massive MIMO systems than conventional MIMO. Solving the pilot contamination problem and deriving performance bounds of massive MIMO systems have spawned an active area of research which is expected to yield new insights into system design in the coming years.

13.4 Small Cells and Big Data

Big Data analytics is one of the new and challenging topics in information technology. Large-scale datasets are generated by many common applications of the present day including telecommunication systems particularly for dense heterogeneous networks. The sheer volume of data often renders traditional data analytics using a central processor and storage impossible. Big datasets are often incomplete, noisy, and data mining has to be performed in a distributed and dynamic fashion to take decisions.

From the perspective of the operator, traditional methods of handling network fault and performance metrics are no longer sufficient in heterogeneous networks [6]. Fast, automated algorithms have become necessary to help the operator in the massive task of network optimization due to the dense deployments of small cells. A dense heterogeneous network with a large and diverse traffic load also generates a huge volume of data in the core network elements. If C-RAN systems are implemented, a huge amount of signal processing and storage is required at the multiple nodes that virtualize the RAN. Some of the tools from Big Data analytics can therefore come in handy. The union of small cells and big data will prove to be an exciting new research field.

13.5 Concluding Remarks

Heterogeneous networks have led to massive improvements in the quality of service offered to the customers of a wireless network. To enable this, a continuous stream of innovations have taken place that encompass fundamental technological research with new business models. This has been enabled by broad-based consensus and cooperation between multiple industry players within the aegis of 3GPP.

The process of innovation is far from over however, and this final chapter has mentioned some new areas of research in heterogeneous networks. The future holds the key to more exciting technological developments, some of which will be adopted in practical systems with relative ease and improve network performance. Some developments may prove to be more challenging to adopt but, by studying the challenges, new insights will be gained. Either way, heterogeneous networks will continue to grow in volume and sophistication in their quest to meet the ever-increasing data demand.

References

[1] Nakamura, T., Nagata, S., Benjebbour, A., Kishiyama, Y., Hai, T., Xiaodong, S., Ning, Y., and Nan, L. (2013) Trends in small cell enhancements in LTE advanced. *IEEE Communications Magazine*, **51**(2), 98–105.

[2] Pi, Z. and Khan, F. (2012) A millimeter wave massive MIMO system for next generation mobile broadband. In *Proceedings of Asilomar Conference on Signals, Systems and Computers*, November 2012.

[3] Fernandes, F., Ashikhmin, A., and Marzetta, T. (2012) Internfernce reduction on cellular networks with large antenna arrays. In *Proceedings of IEEE International Conference on Communications (ICC)*, June 2012.

[4] Hoydis, J. (2013) Massive MIMO and hetnets: benefits and challenges. IMSS, available at http://www.flexible-radio.com/sites/default/files/attachments-11/newcom_presentation.pdf (accessed November 2013).

[5] Hosseini, K., Hoydis, J., Brink, S.T., and Debbah, M. (2013) Massive MIMO and small cells: How to densify heterogeneous networks. In *Proceedings of IEEE International Conference on Communications (ICC)*, June 2013.

[6] Rayal, F. (2013) Are we ready for small cells? Available at https://communities.cisco.com /community/solutions/sp/mobility/blog/2013/07/15/ (accessed November 2013).

Index

Heterogeneous Networks in LTE-Advanced, First Edition. Joydeep Acharya, Long Gao and Sudhanshu Gaur.
© 2014 John Wiley & Sons, Ltd. Published 2014 by John Wiley & Sons, Ltd.
Companion Wesite: www.ltehetnet.com.

Printed in the United States
By Bookmasters